California Drawbridges
(1853-1995)

The Link To California's Maritime Past

by Bernard C. Winn

DO YOU REMEMBER?

**Do you remember the boats on the river
Hugging the fog bound shore
And do you recall the sights and the sounds
As the bridgetender opened the draw**

**Back in the days when wheat was king
And the rivers ran deep and wide
Back in the days when the Delta Queen
Sailed with the evening tide**

**Do you remember old Firebaugh's Ferry
And the route of the Butterfield Stages
And do you recall the ports on the river
Whose names are lost to the ages**

**Back in the days when sail was king
And tall ships filled the bay
Back in the days when wooden scows
Piled their decks with hay**

**Do you remember old Port Costa
And the wharves on Pacheco Creek
And do you recall Old Sacramento
And the mud that they called streets**

**Back in the days when gold was king
And violence filled the land
Back in the days when Colt was law
And ruled with an iron hand**
B. C. W.

This book is dedicated to the memory of Ferne Winn Hickman 1891-1994

Published by Incline Press 2 Townsend Street, 2-213 San Francisco, CA 94107
Library of Congress Card Number: 95-94861
ISBN 0-9615161-8-6

Front cover photogrtaph; Isleton Bridge: California State Library.
Title page photograph; Old Corte Madera highway bridge: CALTRANS

FOREWORD

This book is primarily concerned with the history of California's moveable bridges, past and present. Technical details pertaining to the type, style and mechanics of individual bridges are kept to a minimum.

Contrary to common belief, drawbridges are not limited to the type of structure associated with castle moats. By definition, any bridge with the capability to move a section of its roadway to allow or prevent water vessels from passing safely through is a drawbridge. The moveable section may lift straight up (vertical lift); rotate 180 degrees (swing); float out of the way (pontoon); lift up from one end (bascule, bascule lift); fold away (folding); or slide away horizontally (retractable). Of the above, all except the fold away type are known to have been used in California during its long maritime history.

Unlike fixed or rigid bridges, drawbridges are machines. As such, they require a degree of thought and care. They are complicated by special equipment such as locks, gates, signals, motors and other machinery. Also unlike fixed bridges, they are subject to regulations that apply to both their roadway and the waterway they span.

The state's early bascule bridges were actually very similar to the moat bridges of Medieval Europe. They consisted of little more than a moveable span lifted by cables attached to one or two booms. The mechanism to lift the span was usually a crank or a block and tackle. Counterweights were seldom used on the earliest bridges.

Today, their widely used modern-day counterpart is especially adapted to urban areas where room for a moveable bridge is often limited. They are also well-suited for narrow waterways used by large ships because of their unlimited vertical clearance when open. With machinery and counterweights below street level, some of today's bascule bridges occupy only slightly more space than a fixed bridge of comparable size. Until the advent of the vertical lift, bascule bridges were usually referred to as lift bridges.

Vertical lift bridges are suited for narrow and wide waterways where a single, clear channel is necessary or desired. They generally cause shorter road delays than other types because the vertically moved span often does not have to lift all the way up. However, because they require extra room to provide adequate vertical clearance for large ships, they are less common in urban settings than bascule spans. When the situation calls for a long, moveable span with great strength and rigidity, they are the logical choice. They are particularly popular with railroads for multiple-rail drawbridges over wide channels.

The lift span is supported by cables that pass over pulleys at the top of the lift towers and connect to counterbalance weights. The principle, as it is with a bascule bridge, is to maintain near equilibrium between the weights and the moveable span so only token force is needed to set the span in motion up or down. When closed, the span rests on the lift tower piers.

For about the past 75 years, swing bridges have been the dominate type of moveable bridge in

California. The reason is probably one of economics rather than engineering, however. Although they are usually less expensive to build and maintain than the other two leading types, they do have some serious disadvantages.

Besides using valuable space in the navigation channel, the center pier, which supports the swing span, is very vulnerable to passing ships. Their accidental meeting with ships has been the ruination of many a swing bridge. The necessity of opening the bridge all the way, regardless of the size of the passing vessel, is another undesirable feature. In spite of these drawbacks, however, they are a hearty lot. One such bridge served for 71 years on the busy Oakland Estuary, followed by 17 years at a second location. The Bacon Island Bridge over Middle River is at its third.

While only a few pontoon bridges are recorded in this book, there is little doubt others existed (mostly private). Most pontoon bridges operated by allowing one end of the span to float or be pulled out of the channel. A second type retracts its floating sections back under its approach ramps to open the channel. There has only been one such a bridge in California.

The terms “drawbridge” and “moveable bridge” are synonymous, with one exception. A removable span (barge and boom) bridge, while not a moveable bridge in the literal sense, is sometimes classified as one. When a bridge is required over a navigable waterway but is never expected to open for water traffic, permission can be granted to insert a removable span in place of a moveable span. In the event of a change in the circumstances, or an emergency, the span could be lifted out to provide a clear channel. There are several such bridges in the area covered by this book, but only a few are listed.

While most of California's drawbridges have been located on San Francisco Bay and its tributary rivers, there were, and still are, others. There have been numerous drawbridges in the Humboldt Bay area near the Oregon border and several in Southern California, mainly in Long Beach Harbor.

Until April 1, 1967, the Army Corps of Engineers administered the rules and regulations pertaining to the building, maintenance and operation of drawbridges over the navigable waters of the United States. On that date, the responsibility was passed to the U.S. Coast Guard.

INTRODUCTION

From the 1850s until after the turn of the century, boats of one kind or another navigated the San Joaquin and Sacramento Rivers for approximately 250 miles north and south, into the heart of California's great agricultural valleys.

Those two rivers and their tributaries, along with the Petaluma River, Napa River and numerous other smaller rivers, creeks and sloughs, carried millions of tons of freight and thousands of passengers each year. During those early years, the waterways served as the principal means of transportation, commerce and communication between San Francisco, Sacramento and the hundreds of landings on the waterways.

Until the gold rush of 1849 and for some years to follow, Indian canoes were the primary means of crossing many of the rivers. As the population grew and settlements expanded, bridges were built to get people and goods from one side to the other. Even before the law demanded it, common sense dictated that all bridges across navigable waters had to have a moveable span to allow boats through.

While no one knows for sure just when or where California's first drawbridge was built, the earliest recorded one spanned San Antonio Creek between Oakland and Clinton in 1853. Today, 142 years and 250 plus bridges later, California still has 70 moveable bridges in place. Many of those, however, are no longer subject to opening and several have not opened for water traffic in at least three decades.

With but a few exceptions, river commerce on a large scale is a thing of the past in California. Places such as Hill's Ferry, Alviso, Port Costa, Chico Landing, Pacheco and dozens of other once busy river ports or landings have not seen a steamer in anyone's memory. Today, a decaying or inactive drawbridge standing astride a sluggish stream is the only reminder of the once exciting steamboat days in California.

Historians have been neither generous nor accurate in their writings on drawbridges. The inaccuracies usually involve dates, but sometimes include locations and bridge type. Unless a bridge is demolished by a spectacular explosion or a wayward ship, its demise is seldom mentioned in historical accounts.

While the author has taken great care to present a complete and accurate account of California's drawbridges and their history, he acknowledges that there are undoubtedly some errors and omissions.

Although new drawbridges are still being built in the state, the number of active ones is slowly declining. There is, for example, only one operating drawbridge on the entire San Joaquin River and it has opened only twice in the past eleven years. On the Sacramento River, there are no active bridges above Meridian. On the bright side, one new swing bridge was completed in 1995 and two new drawbridges (one vertical lift and one pedestrian/bike bascule) will be completed in 1996.

ACKNOWLEDGMENTS

I am pleased to give credit to the institutions and individuals who, by their assistance, helped make this book a reality. My sincere appreciation goes to the staff of the Bridge and Structure Division of the California Department of Transportation (CALTRANS). Much of the basic information regarding state and county drawbridges, and many of the photos are from their voluminous files.

This book would also be much less than it is without the valuable information and rare photos and data obtained from the files of the U. S. Coast Guard Offices in San Francisco and Alameda. To Lt. Commander F. Thrall, former Chief, Aids to Navigation, and Wayne Till, currently Chief, Bridge Section, my thanks for the courtesy and help given during the many hours of research in their offices.

For information pertaining to private, county, railroad and extinct bridges of all kinds, I wish to thank the directors and staff of the many county departments of public works; the California Toll Bridge Authority; Southern Pacific, Western Pacific and Santa Fe Railroads.

I would like to acknowledge the valuable help received from the many librarians, clerks, public officials and curators, whose names I never knew.

Special thanks go to my wife, Audrey, for her valuable help and patience during the long process from idea to publication.

Photo credit is given in the caption with the photos. If no credit is given, the photo is from the author's collection.

TABLE OF CONTENTS

1. SAN FRANCISCO	8
2. OAKLAND-ALAMEDA	20
3. SOUTH BAY	35
4. NORTH BAY AND TRIBUTARIES	46
5. SACRAMENTO RIVER	79
6. SACRAMENTO RIVER DELTA	106
7. SAN JOAQUIN RIVER	121
8. SAN JOAQUIN RIVER DELTA	143
9. NORTHERN CALIFORNIA	154
10. SOUTHERN CALIFORNIA	160
BIBLIOGRAPHY	171
INDEX	174
AUTHOR PROFILE	178

CHAPTER 1

SAN FRANCISCO

San Francisco has a long history of drawbridges, beginning as early as 1854. All of its moveable bridges have spanned either Channel Street (formerly Mission Creek), Mission Bay or Islais Creek (formerly DeVerrs Creek).

Mission Bay no longer exists and Channel Street and Islais Creek appear to be waiting for someone to rediscover them. While "The Mission Bay Plan," a plan to develop the area south and east of Channel Street into a huge residential and business community, seems to have been stillborn, it does look like some sort of change is in the wind.

CHANNEL STREET

In 1854, one of California's earliest drawbridges spanned Mission Creek near where, today, Brannan Street intersects Ninth Street. An even earlier drawbridge (pontoon) may have preceded this wooden, hand-operated bascule bridge.

The creek at that time could be navigated by small craft from Mission Bay to the Mission Dolores. A small lake, known as Laguna Dolores, occupied the area now bounded by Guerrero Street and South Van Ness Avenue, between 15th and 19th Streets. The success and popularity of this bridge is best illustrated by a quote from a letter printed in the *Bulletin* dated December 2, 1855.

> There is no country in the world where a man can pay out money and receive no return therefrom with the same facility as in California. Take for example, the opening of the Brannan Street Bridge, which cost property holders...along the route of said road, about $80,000...to build a free road to the Mission. But to cross Mission Creek in a narrow place, required a drawbridge, to accommodate Mr. Dow's distillery. To make a drawbridge to open and shut with a crank, is the simplest thing in all countries except California...True while this bridge had no keeper, it worked well. Recently the Board of Supervisors gave a keeper $1800 a year to take care of this bridge...No sooner was this done, then every Sunday, every holiday, every race day, this same bridge is so shockingly out of repairs as to turn all traffic on the plank road! Gas! Humbug!! Gammon!!....

In 1866-67, the city paid $4,097 for the construction of a new bridge over Mission Creek at Brannan Street. It is not known how long that bridge survived or was active, but an 1891 nautical chart indicated a drawbridge still at that location.

On November 5, 1850, C. L. Wilson was authorized by the City of San Francisco to build the above mentioned plank toll-road from the south end of Kearney Street to the Mission Dolores. The contract was for a seven year period. By July of 1851, omnibuses, operated by Halland Crandall, were running on the plank road. The road, which reportedly cost $96,000, paid its stockholders a return of nearly 8% per month. Tariffs for vehicles and animals on the road were as follows: one animal, 10 cents (25 cents if ridden), vehicle and one animal 50 cents, two animals 75 cents and 25 cents for each additional animal.

Over the years, the mud flats of Mission Bay were filled and an embarcadero with deep-water piers took their place. Laguna Dolores was filled in slowly until it completely disappeared. Mission Creek, through silting and other means, became narrowed to the point where, by the early 1900s, it was confined to an underground culvert above Ninth Street.

Today, Mission Creek (Channel Street) extends from the bay side of Third Street to approximately Seventh Street. Its depth averages 42 feet for approximately 60 feet each side of center to beyond Fourth Street.

To old timers, the "Creek," with its bridges, is best remembered as the location of the "hay wharves" and, later on, the "banana boats." In the 1880s, Channel Street was a shipping point for lumber coming in from up and down the coast. During the period 1880-1890, many of the coastwise lumber schooners themselves were built at several small shipyards on the north side of the creek.

While there have been many changes over the past 140-plus years, it is unlikely Channel Street will suffer the fate of Mission Bay and Laguna Dolores.

Long Bridge * Mile 0.1

Completed in 1867, Long Bridge had moveable spans over both Mission and Islais Creeks. Photo: Bancroft Library.

In 1867, the Bayview and Potrero Railway built a $60,000 trestle bridge (Long Bridge) from the foot of Fourth Street, south to the Bay View Racetrack, near Hunter's Point. The bridge spanned the shallow waters of Mission Bay and the mouth of both Mission and Islais Creeks. Carrying two sets of horsecar tracks, it was built to transport workers to industries south of town and to increase sales of lots in "South San Francisco" near Hunter's Point.

Near the north end of Long Bridge, a wooden, hand-operated swing span was erected over the channel of Mission Creek. That simple structure was the first recorded swing bridge in San Francisco.

Between 1860 and 1885, Southern Pacific managed to buy nearly all the water lots on Mission Bay north of Channel Street between Third and Seventh streets. It also acquired the lots south of Channel Street to Sixteenth Street between Fifth and Seventh. By 1874, Southern Pacific also controlled the Bay View and Potrero Railroad. Other railroads with rights-of-way in the area included the San Francisco and San Joaquin Valley Railroad and the Santa Fe.

During the 1870s and 1880s, Channel Street, between Third and Seventh Streets, became more clearly defined. As Southern Pacific continued to fill in its lots on both sides, wharves were built and piers jutted out into the creek.

Fourth Street * Mile 0.2

In c1875, a manually operated, wrought iron swing bridge was built across Channel Street at the foot of Fourth Street. Carrying the tracks or the Southern and Western Pacific Railroad, it served the city's

The 1892 version of the Fourth Street Bridge over Mission Creek. With some repairs and improvements, it lasted until 1917. In 1901 alone, 6770 ships passed through this bridge. Photo: Roy Graves.

"unofficial dump," located on Southern Pacific property south of the bridge. It unwittingly played a part in the filling-in of Mission Bay.

Short lived, the bridge served for less than 20 years before being replaced. Its replacement, a $26,000 steel swing bridge completed in 1892, was reportedly poorly planned and built without proper foundations. Hit repeatedly by passing ships, it underwent extensive repairs on numerous occasions, including a new swing span in 1907. Although in a state of general disrepair for most of its 25 years, 6770 ships passed through its draw in 1901 alone, the only year for which records are available. Horses, motor vehicles and trains shared the span.

The existing bridge at Fourth Street (The Peter R. Maloney Bridge) is a Strauss designed, bascule span, built by Thompson Bridge Company.

Accepted by the Board of Supervisors on May 23, 1917, the $89,672 bridge began service with a wooden plank roadway, divided down the center by a set of railroad tracks. By 1941, when the bridge was overhauled and redecked with steel, the railroad had stopped using the bridge, so the tracks were removed.

Weighing 951 tons, this true-balanced cantilever bridge now has four traffic lanes and a sidewalk outside each truss. When its single leaf is raised, the bridge provides a 75-foot channel. In the closed position, it is a barrier to just about everything that floats. The leaf is raised by two heavy duty electric motors powered by direct current supplied by PG&E.

The existing Fourth Street Bridge as seen from Blanche's garden. Built in 1917, it is one of three Strauss designed bridges in San Francisco.

Ships drawing up to 20 feet of water and measuring 50 feet in the beam can pass through this bridge. With no adequate docking facilities above Fourth Street, however, it has been a long time since any ship has ventured that far.

In 1985 this bridge was determined eligible for listing in the National Register of Historic Places. The distinction was granted on the basis of its being the only example of this type of bridge in California. The Strauss overhead counterweight was an important innovation in bascule bridge design.

Scheduled for a seismic retrofit in 1995, along with some other repairs and restrengthing, this span may also be raised three feet.

On the southwest bank of Channel Street, adjacent to the bridge, sits a rustic red building that, for 29 years was, the Galerie de Blanche. The lunchtime, seafood eatery had, and still has, a pleasant waterside garden. Blanche no longer owns the restaurant, but she still tends the garden and greets visitors daily. Friend to seamen, dock workers, office staff, and the city's famous and near famous, her visitor's book contains many wonderful sentiments and praises, including one that about says it all: "Blanche - The spirit of the real San Francisco!" Herb Caen 11/6/91.

Roy Graves, remembered by all rail and history buffs, was once a tender on this bridge.

Once one of the busiest spots around the bay, Channel Street is now a quiet backwater. A small marina above this bridge is managed by the Mission Creek Conservancy, a nonprofit cooperative. It maintains the marina and rents docking space to owners of houseboats and other pleasure craft. It also manages the land adjacent to the street above to insure a safe and attractive approach to the marina.

Kevin M. O'Connell, a long-time resident of the marina, tells of the Conservancy's "First Annual Mission Creek Bridge to Bridge Bike Race." With a distance of only 0.2 miles, the fund-raising event was to be more a "happening" than an actual race. Tee-shirts proclaiming "I Survived the 2/10-mile Bridge to Bridge Bike Race," were on sale, as were refreshments and other souvenirs of the event. An error on the part of the San Francisco newspaper billed it as a two-mile race, but once the fun-loving crowd realized the mistake, they took it in stride and had a good time. According to Mr. O'Connell, "Some of the bicyclists solved the problem by riding the course ten times."

"The Second Annual Bridge to Bridge Race" held the following year proved to be the last.

Seventh Street

The City and County of San Francisco built a small swing bridge over the creek at Seventh Street in 1879. The iron structure, which cost the taxpayers $7,187, served for at least 15 years. The need for the bridge ended when a road was built across the creek a few hundred feet above it. The removal date of the bridge is not known, but it was reportedly still in place in 1894. By 1913, the creek no longer reached Seventh Street.

Sixth Street

Built in 1891, this was the second bridge to span the creek at this location. Out of order for most of its last 20 years, it was removed in 1965. Photo: S.F. Public Works Department.

In 1879, the city paid $7,061 for the construction of a 35-ton wrought iron swing bridge over Mission Creek at the foot of Sixth Street. Its manually operated swing span was damaged by passing ships so many times, repairs over the years amounted to several times its original cost. In 1891, after only twelve years of service, this bridge was replaced by a steel, thru truss swing bridge.

Similar in appearance, but much sturdier than its predecessor, the new bridge had a 23-foot roadway and a 122-foot swing span that provided two 50-foot draws.

Manually operated at first, it later acquired a steam engine to open and close the span. At some point, many years after the new bridge had been in use, a steamer sank crosswise in the channel about 300 feet downstream. The creek above the barrier became permanently closed to all but shallow-draft vessels.

Out of order off and on for most of the last twenty years of its existence, the bridge's swing span was closed to road traffic and locked in the open position in June of 1957. In 1965, the city ordered the bridge dismantled and removed from the site.

Third Street

In 1895, the Atchison, Topeka and Santa Fe Railroad (ATSF) selected the area south of Third Street for their San Francisco terminal and got approval to buy the land from the State of California. The railroad filled in the adjacent land on both sides of the channel and built docks along the shoreline.

Railroad Street was extended across their property from the south to where a double-leaf bascule bridge was erected over Channel Street in 1904-1905. Upon completion, the bridge, built jointly by Santa Fe and Southern Pacific Railroads, was turned over to the city. Third Street, at that time, ended at China Basin, an area consisting largely of marshlands through which Mission Creek flowed into the bay.

The Strauss, Page type trunnion bridge had two 56-foot leaves that, when opened, allowed 82 feet of clearance between fenders. The $110,000 span boasted a 37-1/2-foot wide roadway made of laurel blocks. A 10-foot sidewalk ran along the upstream (west side) of the span. In later years, a prominent sign warned, "No vehicle shall be driven across the bridge faster than a walk and no more than one streetcar shall be on the drawbridge at any one time." This rule was overlooked during rush hours when it was not uncommon for as many as 80 streetcars and 200 autos to cross the span each hour. The bridge did not carry streetcar tracks when first opened; they were added several years later.

The two leaves were raised by power from a pair of steam engines which were kept busy with 7000-plus openings a year.

In December 1904, while the bridge was still under construction, an earthquake put a 3-inch crack in the south abutment at the junction of the front wall and the west wing. The result was a settling and slipping of the wing toward the channel. The 1906 quake, coming after the bridge was finished, intensified the movement of the south abutment. A jury-rigged arrangement of struts and timbers between the abutments stabilized the situation and bought some time. It stopped the slippage, but there was still a tendency for the abutment to tip to one side, allowing the south leaf to move forward. The result was a poor fitting south leaf. To improve the fit, an inch or two of the entire width of the span was cut off each year until a total of twelve inches had been removed.

In an effort to end that problem once and for all, a $12,000 repair job was undertaken in 1928. Two massive concrete anchors, installed about 75 feet behind the bridge and in line with the trusses, were tied to the wings on each side. Fortunately, a new bridge was in the works, so the whole "Rube Goldberg" affair had to work for only about three years; which it did.

The original double-leaf bascule bridge built at Third Street in 1904. This 1916 view is from the west. Photo: San Francisco Public Works Dept.

This 1916 view of the original Third Street Bridge over mission Creek is through the spans in the direction of town. Photo: San Francisco public Works Dept.

The existing Third Street Bridge c1933. View is looking south with the 2550-ton moveable leaf in the partly open position. Photo: San Francisco Public Works Department.

By the time the city replaced the old bridge in 1933, it had been obsolete for at least ten years. Third Street, which wasn't even paved in that area until 1917, had long since reached to the southern limits of the city and was a major traffic route.

Opening day for the existing Third Street Bridge, May 12, 1933. The structure has since been renamed the Francis "Lefty" O'Doul Bridge. The busiest of San Francisco's three drawbridges, it opens on an average of 12 times a day during the summer months. Photo: S.F. Public Works Department.

United Fruit Company, with its terminal on Channel Street, wanted to bring in larger ships but couldn't because they wouldn't fit through the bridge. After suffering through many years of the troublesome and limiting drawbridge, it threatened to move elsewhere if the city didn't build a new bridge with more horizontal clearance. In response to that threat and to the obvious need for a new bridge, the city built the existing span.

On the morning of May 12, 1933, a long procession of automobiles, led by the Mayor's chauffeured car, waited on Third Street as the huge span of the new bridge slowly lowered into place across Channel Street. After the ribbon cutting ceremony, with the usual speeches, one of the world's largest single-leaf bascule drawbridges opened for traffic. Hundreds of spectators, two streetcars and a Belt Line Railroad train took part in the dedication ceremony.

The awarding of the contract to Barrett and Hilp caused a storm of protest because their bid of $552,590 was the highest by nearly $33,000. In defending their decision, the city officials argued that the bid was for a superior, Strauss designed and patented bridge. Joseph Strauss was already famous in San Francisco and was the chief engineer on the Golden Gate Bridge which was well into the planning stage.

On April 13, 1933, just a month before the bridge opened, the asphalt covering on the roadway caught fire, causing considerable damage. In another incident, a man was fatally injured by a falling steel beam. As a result of its poor start, it was dubbed the "bad luck bridge."

The cost of the bridge was shared 50/50 by the City of San Francisco and the State Harbor Commission. The Port of San Francisco, under the authority of the State Harbor Commission, owned the Belt Line Railroad which used the Third Street Bridge.

To prevent disruption in traffic on the busy waterway (average of 21 openings per day at that time) the old bridge had to be dismantled and removed while in the open position. Further complicating the issue was the fact that the channel had to be widened from 82-1/2 feet to 103 feet and dredged to 40 feet, all within approximately one month from the start of the demolition.

While blasting the piers and abutments, great care and precision was called for to prevent damage to the nearby Southern Pacific and Santa Fe warehouses. One day, however, after dynamite was loaded into one of the abutments and ready to shoot, the contractor discovered he didn't have a blasting permit. Not wanting to hold up the job, he decided to wait until after midnight, then shoot it anyway. If anyone found out, it would be too late to do anything about it. Apparently no one complained about the lack of a permit, but several broken windows in the Southern Pacific warehouse kept it from being a perfect crime.

As with the dismantling of the old bridge, the huge bascule span of the new bridge had to be fabricated in the open position. During the demolition of the old and construction of the new, all streetcar, railroad and auto traffic was routed over the Fourth Street Bridge.

Weighing in at 2550 tons, the 80-1/2-foot by 135-foot span provides 103-feet of horizontal clearance, compared with only 82 feet for the old bridge. The superstructure contains 800 tons of structural silicon steel; 400 tons of structural carbon steel; 90 tons of cast steel; and 40 tons of machinery.

The leaf is operated by two 200 h.p. electric motors, located above the roadway between the two main trusses. Either motor alone can open the draw in 60 seconds. The controls for the bridge are located in the control house on the north bank. Originally, one man operated the controls from the control house, while a signalman, from his vantage point at the opposite end of the bridge, made sure all traffic was clear of the span before signaling that it could be opened. Today, with the aid of six TV monitors that allow the operator to see every part of the bridge, a signalman is no longer needed. A metal plaque commemorating the bridge's opening is affixed to the sidewalk side of the abandoned signalman's shack on the southeast bank of the channel,

The partially raised Third Street Bridge as seen from the southwest side of Channel Street.

When completed in 1933, the bridge had in the center, a 49-foot roadway, two sets of streetcar tracks, and one set of railroad tracks for the State Belt Line Railroad. Outside of the east truss was a six-foot walk and outside of the west truss, another six-foot walk, plus an 11-foot roadway. In all, there were six traffic lanes.

Steel grating replaced the wooden deck in 1949,

the same year the streetcar tracks were removed. Today, due to the increased size of autos and trucks, there are only four vehicle lanes. The center of the span is divided into two northbound lanes and one southbound. The other southbound lane is located outside the west truss. The sidewalks are still in place, but the railroad tracks have been removed or paved over.

According to recent records this bridge averages nearly a dozen openings a day in the summer and about half that in the winter. Many of the boats using the channel require the bridge to open only a few feet above the roadway, however.

Extensive repair work proposed for 1995 includes raising the entire structure three feet. The project will give the bridge three additional feet of clearance, but the real purpose is to lift the lower chords of the bridge out of the salt water.

In 1990, the city named this span the Francis "Lefty" O'Doul Bridge in honor of the manager of San Francisco's former baseball team, the San Francisco Seals. Listed on the National Registry of Historic places, this bridge also appeared in the James Bond movie, *View to a Kill,* as well as many other lesser known movies.

The San Francisco Belt Line Railroad, which once ran from the Presidio and Fort Mason, along the Embarcadero to a point south of the Third Street Bridge, ceased operation in 1993. Kyle Railroad Co., which specialized in small railroads of that type, was operating the railroad for the port when the end came.

ISLAIS CREEK

Third Street * Mile 0.4

At the turn of the century, Islais Creek, formerly DeVrees Creek, was referred to as Islais Swamp. The natural drainage outlet for a basin occupying nearly 5,000 acres it flowed eastward into a shallow marshy bay before reaching San Francisco Bay.

While Mission Creek and Islais Creek were not officially declared navigable by the state until 1868, a city ordinance passed in 1864 called for moveable spans to be included in all bridge spanning those waterways.

Built for the city of San Francisco by Southern Pacific in 1915, this bridge across Islais Creek cost $214,000. Photo: S.F. Public Works Dept.

Long Bridge, completed in 1867, had a swing span over the navigable channel of Mission Creek and a moveable span of unknown type at Islais Creek.

In 1875, the Southern and Western Pacific Railroad built a swing bridge over Channel Street at Fourth Street and also a drawbridge of some type over Islais Creek at Kentucky. The railroad right of way ran parallel to Long Bridge until a short distance past the mouth of Islais Creek.

By 1885, however, both bridges over Islais Creek were gone and the channel was obstructed by an embankment of rock and earth. A 28-foot culvert in the embankment was the only allowance made for the free flow of water. Built in line with the then Kentucky Street, the "Kentucky Fill" effectively closed the rest of the creek to all water craft.

Although the general consensus at the time was that the creek would never serve a useful purpose, there were those who felt it should be improved. That thinking was evident in c1900, when the Santa Fe Railroad built a trestle on the embankment to carry its rails along Kentucky Street to the area of Mission Creek. In receiving permission to lay the tracks across the waterway, the railroad agreed to add a moveable span, at some later date, if ordered to do so.

In 1906, thousands of tons of debris created by the earthquake were dumped into the tidal marshlands on both sides of the creek, east and west of Kentucky Street. The resulting channel was dredged and later improved by the construction of wharves and piers.

In June of 1914, construction began on the first modern steel drawbridge over Islais Creek. The 1414-ton Strauss trunnion bascule bridge was completed in late 1915. Built by the Southern Pacific Railroad for the City of San Francisco, it cost the railroad about $214,000.

Two sets of train tracks occupied the center of the bridge and two vehicle lanes were located outside the trusses; one on each side. The tracks were shared by the Southern Pacific and Santa Fe Railroads and the Market Street Railway. Motor vehicles were also allowed to use the center of the bridge when not occupied by rail traffic.

In 1918, new wharves were built along the channel, which had been dredged to 24 feet for its nearly 1000-foot length.

Up stream from the bridge a large copra processing plant and the Rosenberg Brothers rice mill were built. East of the bridge was a grain elevator built by the State Harbor Commission and operated by the Islais Creek Grain Terminal Corp. It processed barley for export.

By 1871, San Francisco's "new Butchertown," was well established at Islais Creek. Over 100 buildings and stables were built on piles over Islais Creek, southeast of Kentucky Street. Before the sardines disappeared from the west coast in the 1950s, that same area, south and east of the bridge, was the center of the largest sardine canning industry in the world.

Dedicated on March 4, 1950, the "Levon Hapag Nishkian Bridge" over Islais Creek is the city's newest and most modern drawbridge. It is also the least conspicuous.

In 1940, traffic at the Islais Creek Bridge averaged six trains, four ships and 12,000 autos per day. By 1943, due to the activity at Hunter's Point Shipyard and other war related industries, the auto traffic had more than tripled. In 1942, busses replaced the streetcars on that line. During the Second World War, large seagoing tugs moored above the bridge to make repairs and take on supplies.

Due to steel shortages caused by the war, the new bridge scheduled to be built across Islais Creek in the early 1940s did not become a reality until 1950.

It is unlikely that any other local public work project, costing upwards of $1.3 million, ever had such an ignoble beginning as this: the city's most modern and desperately needed drawbridge. It was an unfortunate situation, but not entirely without cause. At least twice before it actually opened, dedication dates were set and ceremonies planned, only to have them canceled for one reason or another. Finally, on March 4, 1950, the bridge was suddenly ready to open and open it did. Only one city official made it to the opening, and his only official act was to be the first one across the bridge.

The structure is a low profile, double-leaf bascule span that looks nothing like the Third Street Bridge. In fact, unless motorists have had the experience of waiting for the bridge to open or close, they could be unaware that it is a drawbridge. The official name of this span is the "Levon Hapag Nishkian Bridge."

Sixty-eight feet wide by 100-feet long, it carries six traffic lanes and two pedestrian walks. The steel grid roadway is divided in the center by a raised median strip. Railroad tracks, which were so evident on the old bridge, are missing from this one. During the year long construction of the bridge, traffic was routed via Army and Evans Streets.

The grain elevator on the bay side of the bridge is still in place but doesn't appear to be in operation. Above the bridge, there is practically nothing left. The copra plant is gone, the wharf is gone and there is little or no reason for boats to enter the area. Surprisingly, the bridge openings average about six a month, nearly all for testing, but also for an occasional emergency call for a fire boat.

Many of the streets in the Bay View area were changed from their original names. In the case of present day Third Street, it was Kentucky Street from Channel Street to the south side of Islais Creek. Outbound from there, it was Railroad Street to the county line.

Because of the relative rarity of drawbridges in San Francisco, many residents were, and still are unaware of their existence. To enlighten the public, the San Francisco Department of Public Works, in c1985, held what was billed as "Drawbridge Appreciation Days" to show off the city's three moveable bridges. On two successive Sundays the bridges were opened and closed for the benefit of spectators. Bridge engineers were on hand to answer questions regarding the structures and their operation. It reportedly was well received, so might be worth repeating from time to time.

One hour's notice is required for opening each of San Francisco's three remaining drawbridges.

CHAPTER 2

OAKLAND AND ALAMEDA

Nowhere in California have drawbridges been so plentiful or played so important a role as in the Oakland-Alameda area. At least nineteen moveable bridges have spanned San Antonio Creek, its estuary and San Leandro Bay. The state's first recorded drawbridge spanned San Antonio Creek between Oakland and Clinton (East Oakland) in 1853.

Until 1903, Alameda was a peninsula extending northwesterly from San Leandro. In that year the U.S. Corps of Engineers completed construction of a canal from the end of the tidal estuary, near Park Street, to San Leandro Bay. The canal had the effect of making Alameda an island city, but, more importantly, it accomplished its purpose of flushing the estuary of debris and sewage at the head of the basin. The flushing action also ended a serious health problem and stopped the silting-in of the estuary which threatened its usefulness as a harbor.

San Antonio Creek flowed westerly from the Oakland hills, emptying into the estuary near the present location of Laney College. In the 1850s and 1860s, schooners navigated the creek to load lumber and hides at the embarcadero on the north shore of Lake Merritt.

In 1886, the natural low water depth at the entrance to the estuary was only two feet. The Corps of Engineers, between 1886 and 1898, dredged the waterway to a width of 300-feet and a depth of 18 to 20 feet. Vessels drawing 19 feet could then pass from San Francisco Bay to Oakland at ordinary high water.

Twelfth Street, Oakland

In 1852, Contra Costa County contracted with Mr. T.C. Gilman to build a drawbridge across Indian Creek (San Antonio Creek) at Twelfth Street, between Oakland and Clinton. Oakland was part of Contra Costa County at that time. The contract, in the amount of $7,400, was to be paid in full upon completion of the bridge. Interest, at 5% per month, would be charged on any unpaid balance.

The span was accepted by the Board of Supervisors on March 8, 1853. However, because the county lacked funds at the time to pay the entire amount, final payment was deferred. Considerable time elapsed before Mr. Gilman mentioned the debt owed to him, but when he did, it was in the form of a law suit against the county. He demanded immediate payment of the balance and threatened to force the sale of the courthouse if necessary. Only intervention by the State Supreme Court prevented this action. In 1862 the county initiated a property assessment on the City of Oakland to raise the needed money. The debt, which had grown to $31,611 by this time, was finally paid in full on June 19, 1862.

Meanwhile, in 1855 the county entered into a contract with a Mr. Carpenter to do some needed work on the bridge. In exchange for this work, Mr. Carpenter received permission to collect tolls for a period of one year, at which time the bridge was to revert back to county control. That arrangement did not work out well either, so, after much wrangling, the supervisors paid Mr. Carpenter $6000 just to get rid of him. The bridge went back to being a toll-free county road.

This expensive and troublesome bridge, which consisted of little more than a long wooden trestle with a hand-operated bascule drawspan, is the earliest recorded drawbridge in California.

In 1865, the San Francisco-Oakland Railroad Company built a drawbridge of undetermined type across San Antonio Creek at Seventh Street between Third and Fourth Avenues.

Four years later, in 1869, the Western Pacific Railroad received permission to construct a fixed railroad trestle across the mouth of the creek. For most vessels, that was the end to navigation on San Antonio Creek. It is not known just when the drawbridges were removed.

Webster Street * Mile 5.0

The first of four bridges to span the Oakland Estuary at Webster Street consisted of a long narrow timber trestle with a 200-foot swing span near its center. Built by the county in 1871, it was without railroad tracks or pedestrian walks and barely wide enough for carriages to pass.

In 1880, the South Pacific Coast Railroad, operators of a ferry between Oakland and Alameda, obtained the rights to the bridge and immediately tore it down to make room for a combination railroad and vehicle bridge. It consisted of a manually operated wrought iron swing span and wooden approach spans. An unusual feature on the swing span was its three trusses; the extra one, in addition to adding strength, separated the two vehicle lanes from the narrow gauge railroad tracks. Built for a mere $25,000, it opened for business on May 30, 1881.

The victim of many a wayward ship, it received its share of damaged pilings or worse. On May 30, ~~1900~~ 1890, nine years to the day after it opened, it was the scene of a tragic train accident. According to witnesses of the event, the steamer *Juanita* signaled to pass through and the bridge opened promptly. At the same time, a train coming from Alameda failed to heed the flag signal indicating the bridge was open. Without warning, the engine plunged into the water carrying one passenger car with it. Thirteen lives were lost in the accident. Newspapers reported that if it were not for the many people who immediately came to the rescue, the death toll would have been much higher. Charged with manslaughter, the train's engineer conveniently left the area soon after the accident and the case never went to trial.

This sketch is of the May 30, 1990 accident on the approach span of the 2nd Webster Street Bridge.Thirteen people were killed. The moveable span of this bridge was moved to Bay Farm Island in 1896. Photo: Alameda Library.

Struck by the steamer *Landcaster*, the swing span of the third Webster Street Bridge was a total loss. Part of the steel from the span was melted and recycled into the Posey Tube. Photo: Alameda Library.

In the photo to the right, the last Webster Street Bridge is seen from the Alameda shore. Photo: U.S.Coast Guard

On May 20, 1896, the bridge was declared an obstruction to navigation by the Corps of Engineers. The owners were ordered to remove the bridge or enlarge the draw to 150 feet and provide mechanical means for opening the bridge. So, after 17 hard years at Webster Street, the bridge was dismantled and the swing span moved to a new location between Alameda and Bay Farm Island, where it served until 1953. Alameda was without a bridge at Webster Street for approximately two years.

On February 3, 1900, the third bridge to span the estuary at that location was welcomed by the residents of both Oakland and Alameda. The all steel, swing bridge, with a 40 h.p. motor, became the first bridge on the estuary with an electrically operated draw. It was also possible to open it manually and on at least one occasion it was opened and closed with the aid of a team of horses.

Its wooden roadway carried a single set of tracks for streetcars and trains and two lanes for vehicle traffic. Soon after it opened it suffered a rash of minor ship collisions, a fire and an attempted bombing. Things got better over the next few years, but not much.

On January 7, 1926, the steamer, *Lancaster*, under tow of the tug *Sea Lion*, rammed the 350-foot swing span knocking it off its center pier and into the estuary. With the span damaged beyond repair, the residents of Alameda were once again without a bridge at Webster Street.

Because a vehicular tube being built under the estuary was not scheduled for completion until 1928, the moveable span was replaced and the rest of the bridge repaired. A large portion of the steel from the 750-ton wrecked span was recovered and, after remelting, was used for concrete reinforcement in the Posey Tube. So, even today, part of the third Webster Street Bridge still crosses the Oakland Estuary.

Found guilty of negligence, the owners of the *Lancaster* and the towing company paid the county $69,288 for damages. The original cost of the bridge was $99,535.

On January 17, 1926, H.H. Tillman was awarded the contract to build the new swing span, which was to be an exact replica of the former one. Working against a seven-month deadline, the socially prominent Mr. Tillman even labored alongside his employees on occasion to insure finishing the job on schedule and, incidentally, to avoid a $500 per day penalty if it was not.

The new swing span and the rest of the repairs were completed on time and, on August 17, 1926, the fourth and last Webster Street Bridge was dedicated amidst a celebration the likes of which had never been seen before. A "Queen" reigned during the celebration as people danced in the streets to a half dozen bands. Happy Webster Street merchants provided free refreshments and wide smiles.

Although the center pier and swing span were most of what was replaced, it was, for all practical purposes, a new bridge when it reopened eight months after the accident. The final price tag of $134,000 certainly reflected that fact.

The life of the new bridge, while short, was not uneventful. The captain of the S.S. *Diane Dollar,* eastbound in the estuary on May 21, 1927, either overestimated the speed at which the bridge could be opened or underestimated his own speed. In any event, his ship rammed the bridge and caused extensive damage to the trestle section and minor damage to his vessel. Both rail and vehicle traffic on the bridge was suspended for more than a month. Foot traffic, however, was not effected. People had only to walk the length of the bridge, plus the distance between streetcar connections at each end.

In 1927, the only year for which accurate figures are available, the bridge opened for water traffic 18,756 times, or on the average of more than 50 times during each 24 hour period. Little wonder advocates of the Posey Tube found many supporters.

Soon after the tube opened in September of 1928, the county put the bridge up for auction. Because of its close proximity to the tube, the bid announcement stipulated that the bridge be dismantled and removed without the use of explosives. Duncanson Harelson Company purchased it for $3,100 and, after dismantling it, towed the sections away on barges to a storage area near Antioch.

In 1932, the new owners had the swing span towed to Jiboom Street in Sacramento where it was installed as part of the "Garden Highway Bridge over the American River." It is not recorded whether it ever opened for water traffic at that location.

Webster Street * Mile 5.0

Due to an unusual situation brought about by the building of the second tube under the Estuary, a fifth drawbridge was built at Webster Street in 1960. Unlike the other four bridges, it was a vertical lift bridge and was built parallel to the estuary, rather than across it.

To facilitate the building of the tube and portal on the Alameda side, a 40-foot-deep trench had to be dug from the estuary to near Marina Village Parkway. The mud bottom on the Alameda side proved to be unsatisfactory and had to be dredged to provide the necessary stable underwater base for the tube.

Unable to eliminate the Southern Pacific tracks that crossed the soon to be dredged area, a temporary vertical lift drawbridge with a 90-foot lift span was erected. The structure, which spanned the 450-foot trench, was opened to rail traffic in April of 1960.

It is not known whether the bridge had to open often for water traffic or was maintained in the open position. It was probably the latter, however, because water traffic would likely have been heavier than rail traffic during the construction. Also, when closed, the bridge offered little vertical clearance except at low tide. On one occasion, the lift span and the bridge's entire longitudinal top bracing had to be removed by a floating crane to allow a huge dredge into the area between the bridge and the Alameda portal building.

The second tube under the estuary was dedicated on February 13, 1963. It is not known at what point in the construction the trench was filled in, and/or the bridge removed.

Alice Street * Mile 5.2

Built in 1873 by the Central Pacific Railroad, this wooden trestle bridge, with a manually operated swing span, was in place for 27 years. Removed in 1900, it was replaced by a steel swing bridge completed at Harrison Street in 1898. Photo: Oakland Library.

The first direct railroad service between Oakland and Alameda began on September 29, 1873 with the opening of the Alice Street Bridge. Until that time, train service between the two cities involved a rather long and roundabout trip through Oakland and San Leandro.

Built by the Central Pacific Railroad and used exclusively by their trains, the bridge had wooden approach spans and a 200-foot, hand-operated swing span. In 1896, the Corps of Engineers informed Southern Pacific (the new owners) the bridge would have to be altered to provide a 150-foot clear channel, and, furthermore, its draw could no longer be manually operated. They were given 15 months to comply. Instead of rebuilding, however, the railroad made a deal to replace the Alice Street bridge with a new, larger, and more modern one to be built at Harrison Street. After approximately 25 years of very active service, the decrepit Alice Street bridge was removed in May of 1900.

Harrison Street * Mile 5.1

Southern Pacific completed their new, thru truss steel swing bridge across the estuary at Harrison Street in 1898. The bridge's swing span looked to be a carbon copy of the then existing Park and High Street bridges.

When the bridges at Webster Street and Harrison Streets were open st the same time, less than

140 yards separated the ends of their moveable spans. Ships passing through both bridges had to negotiate a 150-foot wide channel for 800 feet. With the strong tidal currents in the estuary, ships and especially tugs with tows, often had a difficult time of it.

With the bridges opening as many as 100 times (during the busiest nine hours of each day), accidents of one kind or another were common. The worst accident occurred on July 4, 1909. Two commuter trains, each jammed with holiday passengers, collided on the bridge, killing 11 persons and injuring many more. The trains were on the route between the Alameda Mole and Franklin Street in Oakland. Another incident worthy of note happened on January 6, 1920. The Schooner *Tamalpias* signaled for the bridge to open, but failed to stand back and wait. With the draw only slightly opened, the ship's bow struck one end of the swing span, slid under it and stuck tight. The accident happened on the flood tide, so until the tide reached its peak sometime later, it was a tossup as to whether the ship's bow would be forced under water or the span lifted off its center pier. Fortunately, neither happened. With the aid of a floating crane and the ebb tide, the two were separated with only minor damage to each.

By 1915, shipping in Oakland's Inner Harbor was suffering from the congestion the two bridges wee causing. Maritime industries using the waterway were demanding some action by the county.

In January of 1916, Contra Costa County* Southern Pacific Company and an unnamed street railway, agreed to share the cost of a bascule bridge to replace both the Webster and Harrison Street bridges. The county would pay $900,000 of the estimated $2,000,000 cost of the bridge, with Southern Pacific and the street railway paying the balance. Engineering plans for the bridge were approved by the county supervisors in 1917 and bids were prepared. Then, in july of 1918, all the parties involved in the bridge planning were informed by the War Department that the bridge would have to be put on hold until the end of the war. Although the war ended that same year, the bascule bridge never became a reality. Soon after the War Department decision was handed down, both bridges underwent needed repairs to allow them to survive a few more years. *Oakland and Alameda were in Contra Costa County at that time.

The Harrison Street Bridge shown shortly after its completion in 1898. Built to take the place of the Alice Street span, it was a fixture on the estuary until 1924. The Posey Tube now occupies the site. Photo: Alameda Library.

Fruitvale Avenue * Mile 7.7

A legend in its own time, the Fruitvale Bridge earned the nickname "The Broken Bridge," given to it several months before it officially opened in 1902. Constructed for the Corps of Engineers by Darby Langdon & Company of San Francisco, it cost the government $53,759. It was actually completed in 1901, but the machinery used to operate the drawspan would not work. Its dedication was delayed until 1902, when the Corps of Engineers finally accepted it. That became the pattern for the next 70 years.

Taken c1903, this photo shows a train approaching the swing span of the original Fruitvale Bridge. A legend in its own time, this bridge survived for 70 years.
Photo: Alameda Library.

By 1919, there was serious talk of condemning the bridge because of its poor condition. Engineers estimated it would cost upward of $10,000, just to make it reasonably safe again. The county, which operated and maintained the span at the time, made the necessary repairs, rather than build a new bridge. That was only the beginning, however; good money followed bad every few years until the bridge had been paid for twice over. Part of the bridge's problems stemmed from the fact that one side of the swing span was 5.5 feet longer than the other. its three

Originally used by trains, trucks and autos, it was down to just autos when this photo was taken in 1972. Train tracks formerly occupied the area to the right of the center truss. The only major change in the bridge's appearance over the years was the relocation of the operator's house to the area between the trusses above the roadway.

truss arrangement, which was not unlike that on the second Webster Street Bridge, no doubt also added greatly to the span's problems.

Originally used almost exclusively by trains, with the roadway getting only moderate use, the situation was later reversed as autos became more numerous. In 1941, the last passenger train crossed the bridge, and by 1951 all rail traffic on the bridge had ceased and the tracks were removed. In 1968, trucks were banned on the narrow roadway, which was essentially unchanged from the day the bridge opened. There were those who predicted that autos would be the next to go, making the Fruitvale Bridge the only, "pedestrian drawbridge" in the state.

For the first ten years of its existence, while the bridge was manually operated, the bridgetender's house stood at one end of the bridge. In 1911, however, the bridge was electrified and the tender's house, with the controls for opening the span, was raised to the center of the span above the roadway.

During most of the 70-plus years the Fruitvale Bridge obstructed water traffic on the Oakland Estuary, practically everything imaginable happened to it except the one thing everyone really wanted, its removal.

There were periods when the draw was stuck open or closed, or otherwise out of order, for more days during a month than it was in working condition. The bridge, or its protective pilings, were struck innumerable times and by every type of watercraft: including ships, tugs, barges and even a submarine. More than once, its trusses were damaged by autos and, on one occasion in 1937, a freight train derailed on the bridge, sending two cars through the trusses and almost into the water.

This bridge and the High and Park Street bridges were built by the U. S. Government in exchange for permission and rights-of-way to dredge the channel between San Antonio Creek and San Leandro Bay.

In another exchange, for some rights-of-way in Alameda that were owned by Southern Pacific, the government agreed to make the Fruitvale Bridge a combination auto and railroad span. They further agreed to maintain it indefinitely, without cost to the railroad. The county, wishing to lease some government land on the newly created canal, began negotiating in 1909 to take over control and maintenance of the three bridges in exchange for the leases. On November 10, 1913, the Board of Supervisors finally signed the agreement and became the new operators of the bridges.

Everything went along fine until 1939, when the county sent Southern Pacific a bill for $3.24 to cover the cost of some nuts and bolts required for the railroad portion of the Fruitvale Bridge. Southern Pacific naturally refused to pay the bill and suggested the supervisors read the fine print on their contract, which granted the railroad "free" passage forever. The supervisors were outraged. It might be all right for the Federal Government, or even previous county supervisors, to put out tax money for private interests, but not them. Into court it went. After three years of expensive litigation, the famous "Nuts and Bolts Case" was settled. The federal government took back control of the Fruitvale Bridge and the county kept the Park and High Street bridges.

In January of 1972 the government finally removed the ancient Fruitvale Bridge to make room for a long overdue replacement. The new structure, a single-leaf bascule bridge, was built at a cost of approximately $4,000,000. The moveable span is located on the Alameda side and allowances have been made to add a second leaf if necessary. With 15-feet of vertical clearance compared to six on the old bridge, this bridge will not be required to open nearly as often as its predecessor.

Unfortunately, the Fruitvale Bridge's history of mishaps was passed along to the new structure. Actually completed in 1973, its bascule leaf stuck in the open position during its trial run. Months of work and thousands of dollars later, the Bridge was finally opened to traffic the following year.

Trouble struck again on March 29, 1990 when the span was struck by a barge being pushed by the tugboat *Ellen*. The barge, loaded with sand and bound for the Tidewater Sand and Gravel Co. dock, radioed the tender to open the bridge. A malfunction prevented the span from lifting immediately and the tender radioed the tug to stop. The tugboat skipper, Roger Campbell, put the tug in reverse, but a strong tide pulled it into and partly under the bridge before it could be opened. When the impact occurred, the boat was moving at about one knot. Structural damage to the bridge was not major, but it was closed to both vehicle and boat traffic for several days.

The "Miller Sweeney Bridge," the existing bridge at Fruitvale Avenue, was completed in 1973.

With four lanes for vehicles and 15-feet of vertical clearance for boats, there's no comparison as far as most are concerned

While this span was under construction, and traffic was routed over the other bridges, people were heard to ask, "where is that wreck of a bridge, now that we really need it?"

The bridge is named for George P. Miller and Leland W. Sweeney, the men responsible for getting the bridge off the drawing board and across the estuary.

(So.Pacific.) Fruitvale Avenue * Mile 7.7

In 1951 the Corps of Engineers erected a new vertical lift bridge adjacent to the old Fruitvale Bridge for the exclusive use of the railroad. It provides 13-feet of vertical clearance when closed, but its 215-foot long lift span is kept open and lowered only when needed by trains.

It was believed by many that this bridge was built mostly to get even with the county for going back on their deal of 1913.

In 1951, The Corps of Engineers built this vertical lift bridge for Southern Pacific to get trains off the old Fruitvale Bridge. The old bridge is shown in the background.

Park Street * Mile 7.3

Of the original three government-built bridges on the estuary, this was the first to be completed. Carrying a single set of streetcar tracks down the center of its wooden deck, it had a wrought iron, thru truss swing span and wooden trestle approach spans. The price tag of this simple structure was $50,000.

Although finished and scheduled to open on December 7, 1891, it was early 1892 before it actually went into service.

The bridge was constructed just beyond the head of navigation over what was, at the time, best described as a swamp. It was thought that the channel under it would be finished within a year of two but due to several lengthy interruptions, it was actually ten years.

This lightly built bridge was in place for 41 years and was very popular as a fishing bridge. Traffic on the span eventually became so heavy, fishing from the span had to be stopped for safety reasons. "Riding the bridge," as it opened for water traffic, was also a favorite pastime of pedestrians, but that too had to be discontinued for the same reason.

The old Park Street Bridge was the first of the original government-built bridges to span the Oakland Estuary. Completed in 1891, it stood over a swamp for ten years before the excavating project was finished. Photo: Oakland Public Library.

Although enjoyed by the general public, and especially fishermen, the extremely busy bridge was in poor shape for many years before it was retired. Dismantled by Thomas Rigging Company, it was hauled away by Harbor Tug and Barge Company on February 20, 1934.

October 5, 1935, witnessed the opening of the first of three modern bascule drawbridges to be built between Oakland and Alameda. The highlight of the Grand Opening celebration was the public wedding of an Oakland man and an Alameda woman at the center of the span. The ceremony was by way of symbolizing the unity of the two cities joined by the bridge. The wedding kicked off the gala celebration planned for the new four-lane bridge.

The existing Park Street Bridge, was the first of the three bascule bridges now spanning the estuary.

Unlike bridges in some outlying areas, the estuary spans are part of the daily lives of thousands of people and are not taken lightly. After nearly a dozen drawbridges, Alamedans can claim an expertise at celebrating bridge openings; something they never seem to tire of.

The double-leaf bascule span cost approximately $850,000 and has more than proven its worth. With its machinery and counterweights out of view, this 60-year old bridge still pleases the eye.

With four 75 h.p. motors, one operator can raise or lower both 120-foot leaves in minutes. The leaves can be opened individually or in unison, depending on the need.

The span has had its share of accidents over the years, both with autos and ships. On August 18, 1937, for example, an automobile, stopped in the wrong place, was crushed by one of the bridge's huge leaves as it opened to allow a boat to pass through. The accident left the bridge relatively undamaged, but destroyed the car. The driver, who was standing near his car when it happened, was not injured. By coincidence, on the same day, a train derailed on the old Fruitvale Bridge.

High Street * Mile 8.1

The second of the three government bridges was the iron swing bridge completed at High Street in early 1901. As with the Fruitvale Bridge, it too stood astride an unfinished channel. Built by the California

The High Street Bridge c1930. Plagued with troubles from the beginning, it was condemned and closed for several months in 1927. It survived until 1938, however, when it was replaced by the existing High Street Bridge. Photo: Oakland Public Library

Bridge Company in the style of the Harrison and Park Street bridges, it cost the government $24,747.

On May 21, 1901, a fire destroyed the swing span and part of the approaches. Rebuilt, it managed to survive until 1927, when it was condemned and closed. Restored the following year, it reopened to try again. Although far from being the busiest bridge on the estuary, it suffered more than its share of abuse. Old-timers claimed it was news when a ship or barge made it all the way through the draw without hitting at least part of the bridge or fenders. It was second only to the Fruitvale Bridge when it came to problems.

On numerous occasions, it was closed to water traffic, vehicle traffic or both, for anywhere from a few hours to several days. It closed to vehicle traffic for the last time early in 1938 and was removed on September 7 of that year.

Although the existing High Street Bridge resembles its neighbor at Park Street, it has only two traffic lanes, compared to four on the Park Street span.

On December 9, 1939, Alamedans gathered to witness the opening of a new $750,000 double-leaf bascule bridge at High Street. The celebration that accompanied its dedication lacked the usual enthusiasm, however, because many residents were unhappy with their new bridge.

The span had only two lanes, whereas the Park Street Bridge, which cost only a few dollars more, had four lanes. Most everyone agreed it was an improvement over the decrepit swing span, but felt they had been short changed. With the Fruitvale Bridge out of commission most of the time, and a large increase in road traffic, the High Street Bridge soon reached gridlock stage.

Eventually, due to fewer openings and the smoother roadway, the initial disappointment was mostly forgotten and most were grateful for the change. There was also no denying the new structure was a great improvement in reliability alone.

Since the existing Fruitvale Bridge opened in 1970, there have been relatively few bridge problems on the waterway. All three vehicle drawbridges on the estuary are tended 24 hours a day. They are, however, exempt from opening during weekday rush hours (7:30-8:30 a.m. and 3:45-4:45 p.m.

SAN LEANDRO BAY

Bay Farm Island

About the time Mr. Gilman was building the 12th Street Bridge over in Oakland, Messrs. Chipman and Aughinbaugh were making plans for a toll drawbridge between Alameda and Bay Farm Island. A proposed land speculation deal on Bay Farm Island prompted their interest in the bridge project, which cost them nearly $20,000 by the time it was completed in 1854.

Built entirely of wood, the bridge was surfaced with oyster shells, as were the roads at each end. The land deal and the bridge both failed financially, so the owners stopped collecting tolls and eventually abandoned the bridge altogether. After several years of neglect, it rotted to the water line. By 1860 nothing remained but some rotted pilings.

The County of Alameda tried its hand at bridge building in 1874 at the same location, but it too proved unsuccessful. Although dedicated and opened with a wild celebration, the 754-foot long bridge was short lived. The county was clever enough to remove it before it rotted, however, and parts of it were later used for a wharf in Alameda.

A Captain Anderson, who regularly hauled vegetables between Bay Farm Island and San Francisco on his boat, *La Bonita*, claimed that sea lions were so numerous in the area that at times, it was impossible for a person to cross the bridge due to the large number of animals lying on it.

The third bridge, and first successful one to span the channel between Alameda and Bay Farm Island, was constructed around a hand-me-down swing span from the Oakland Estuary. Built in 1881, the swing span was part of the second Webster Street Bridge which was declared a menace to navigation in 1896 and ordered removed.

In 1898 the rickety swing span was barged to its new home and lifted onto the waiting center pier. The fixed spans and approaches at each end were also readied for its arrival at a cost of nearly $20,000.
Due to its eternally loose deck planks and a troublesome center dividing truss, it proved to be a constant source of frayed nerves and accidents , just as it had at Webster Street. Speed and load limits were barely above what one might expect on a foot bridge.

When replaced in 1953 as the oldest drawbridge in California, it was, according to the California Highways and Public Works magazine, "The State's number one candidate for the junk heap." It reportedly opened for water traffic more times in its 15 years at Webster Street than during its 54 years at Bay Farm Island.

The swing span shown here was built in 1881 as part of the second Webster Street Bridge. Moved to this location in 1898, it was in use until 1953. Photo: CALTRANS.

The present Bay Farm Island Bridge consists primarily of many concrete approach spans, and a 125-foot bascule span with a 1,100-ton counterweight. Suspended below the roadway, the counterweight sinks into a well in the pier when the span is raised. The single span offers 92-feet of horizontal clearance and allowances have been made to add a second leaf if ever needed. The cofferdam that makes up the main pier containing the machinery and counterweight was the largest built on a state highway since the construction of the San Francisco-Oakland Bay Bridge. Cost of the bridge and approaches came to $2.3 million. Contractors were Duncason-Harleson Company and Stolte Inc.

The 53-foot roadway carries four lanes of traffic separated by a six-foot divider strip. Opened to traffic on July 1, 1953 and dedicated on July 2, it was, in the words of one of the thousands of persons at the dedication, "Too long a coming."

This bridge is tended from 8 a.m. to 5 p.m. Daily with 12 hours notice required for opening at other times.

This bascule bridge at Bay Farm Island was completed in 1953 at a cost of $2.3 Million. The first modern bascule span in the area with only one leaf, it can be converted to a double-leaf bridge if necessary.

Bike Bridge

This may not be the only bicycle/pedestrian drawbridge in the country, but it certainly will be the only one in California. Started in 1993, this $3,000,000 structure is scheduled for completion in late 1995.

Crossing the channel at a slight angle, it is situated a few yards east of the existing highway bridge. Its 10.75-foot, roadway will carry bikes, pedestrians and vehicles for the handicapped. Connecting Fernside Boulevard shoreline path with the bike paths on Bay Farm Island, it is the final link in Alameda's bike master plan and ABAG'S Trail Around the Bay. Speed limit on the bridge will be 20 mph. No motorized vehicles, except wheelchairs etc., will be allowed. On each side of the bridge is a 4 1/2-foot high railing.

While the hydraulically lifted, 128-foot bascule leaf can be operated independent of the adjacent vehicle bridge, in actual use it acts as a slave to the larger span. When the leaf on the existing bridge opens it will automatically open the one on the bike bridge. The same is true of the warning gates, bells and lights. The two bridges are connected by a small bridge, on the Alameda side, to aid in the dual operation.

Shown under construction, this unique (in California) drawbridge will be completed in late 1995. It is the first and only drawbridge built in California for the exclusive use of bicycles, pedestrians and wheelchairs. In the photo the Bay Farm Island bascule bridge can be seen to the left.

CHAPTER 3

SOUTHERN SAN FRANCISCO BAY AND TRIBUTARIES

REDWOOD CREEK

Redwood City

In the early 1850s, Redwood City had a dozen or more sawmills sending redwood lumber to its embarcadero for shipment to San Francisco. Records for the summer of 1862 show an average of 24 ships a week arriving at the port to load lumber, firewood and shingles. Many logs were also towed from there to San Francisco and other bay points. In 1892, the U.S. Corps of Engineers dredged Redwood Creek to a width of 50 feet for 6,000 feet above the mouth and to a depth of three feet at MLW (mean low water).

Redwood Creek, upstream from the present Port of Redwood City, is now partially contained in concrete culverts under the streets. The port is situated on what is now referred to as Redwood Creek Channel. A federal navigation project, it is maintainedto a 30-foot depth at MLLW (mean low low water) from the bay to the upper turning basin. The only deep-water port on the bay south of San Francisco, it is called on by upward of 250 commercial vessels each year. Lumber is still one of Redwood City's major exports, but recycled metal products, dry bulk, liquid bulk and specialized cargoes are also important. In 1994, specialized small cruise ships carrying nearly 1,100 passengers visited the Port of Redwood City.

The first drawbridge in the south bay area was a hand operated wooden structure erected across Redwood Embarcadero Creek in 1854. The creek passed through town in the area between First Street (now Jefferson) and Main Street. The bridge spanned the creek at "A" Street (now Broadway). Broadway, between Main Street and Redwood Creek, was later renamed Bridge Street.

Redwood City, at that time, was still in San Francisco County. San Mateo County was not established until 1856. The bridge, built by Mr. Littlejohn at a cost of $500, was paid for by the citizens of the town. They tried in vain to get the San Francisco Board of Supervisors to pay for it, and had only slightly better luck with the San Mateo Supervisors in later years.

A concrete fixed span was built at the above location in 1905 to replace a "wooden bridge." It is not known whether that wooden bridge was the original drawbridge or a later one.

In 1868 a new lift bridge (bascule) was constructed over the creek at "B" Street (now Marshall). That bridge remained active until 1884, when flood gates were built across the waterway, keeping all shipping below that point. Shipbuilding activities formerly located above the bridge, moved downstream and the last ship of any size built in Redwood City was the *Perseverance*, completed in 1883.

Taken in c1925, this photo shows the bascule bridge that spanned Redwood Creek at East Bradford Street. Built in 1910, it was the last of Redwood City's three drawbridges. The boat in the photo, the *Frolic*, belonged to the Hanson family, owners of Hanson's Lumber Company. Photo: Redwood City Library.

Redwood City's last drawbridge was completed and accepted by the city on November 4, 1910. A hand operated, wooden structure, it stood on East Bradford, (formerly "C" Street), and provided a 30-foot clear channel. The exact date of its demise is not recorded, but on March 7, 1930, the city asked the War Department for permission to replace it with a fixed bridge.

During the creeks busiest years, vessels drawing up to seven feet of water could reach the "A" Street bridge at high tide. Most of the lumber ships loaded downstream from the bridges.

SAN FRANCISCO BAY

(So. Pacific.) Dumbarton Point * Mile 32.5

Although no longer in use, the Southern Pacific swing bridge is still in place after 85 years. Built in 1910, it was the first to span San Francisco Bay. Photo: Bancroft Library.

The first bridge to cross San Francisco Bay proper was the $1,500,000 railroad swing bridge built by Southern Pacific Railroad in 1910. Presently one of the oldest bridges in the state, it spans the bay between Dumbarton Point in Alameda County, and Redwood City in San Mateo County.

In the 1850s, Dumbarton Landing, later changed to Dumbarton Point, served hundreds of bay scows carrying lumber, hay, and fresh produce to and from landings all around the bay. The road between Dumbarton Landing and Mission San Jose is reportedly the oldest in Alameda County.

Dedicated on September 24, 1910, the bridge did wonders for the City of Newark at the eastern end of the bridge. The bridge's opening was celebrated by a weekend celebration that included a grand ball in the Newark Pavilion; a championship baseball game between Redwood City and Newark; a free-for-all fight and other festivities. All in all, it was a weekend that, according to newspaper reports, "Drained every keg of steam beer in the township." Also on opening day, a passenger train made a special trip across the bridge and back. That trip stands as the only scheduled passenger train to ever cross the Dumbarton Bridge.

Southern Pacific was only one of three railroads wanting to build a bridge across the south bay. The Western Pacific and Santa Fe Railroads also had plans for such a span. Unfortunately, the "Espee" beat them to it.

Taken in 1912, this photo shows a west bound train on the fixed section of the Dumbarton Bridge. Photo: Bancroft Library.

Allegedly, early company propaganda led the public to believe the bridge, billed sometimes as the Dumbarton "Union" Bridge, would actually be a tri-company bridge, shared by all three railroads. When completed, however, the word "union" was nowhere to be found in the name of the bridge, which was to be used exclusively by Southern Pacific.

The only swing bridge to ever span San Francisco Bay, it was referred to for most of its life as the Dumbarton Point Drawbridge.

The seven, thru truss spans that make up the 7,559-foot structure were fabricated in Oakland and hauled down to the bridge site on the old Southern Pacific bay-ferry *Thoroughfare*. Cut in half and shored up at the cut ends, the ferry was transformed into two barges capable of supporting the heavy sections. Tidal action provided the lift to put them into place.

Originally equipped with a steam engine for operating the 310-foot swing span, the bridge was later changed to diesel and eventually to electricity. During the early years, a tug stood by at the bridge during heavy weather or difficult tides to assist ships through the draw.

The 85-year old bridge has not been used by Southern Pacific since before 1982 when it was abandoned and its swing span welded in the open position. Much of the water traffic through the draw is made up of pleasure boats and with the bridge already open, there is no way to record how many there are. In earlier years, the bridge reportedly opened as many as 1,000 times a year.

With gridlock on many highways, several groups around the bay are studying the possibility of using the old bridge to facilitate a commuter rail service between Newark and the San Francisco Peninsula.

(Hwy. 84) Dumbarton Point * Mile 32.0

The first highway bridge to cross San Francisco Bay and, at that time, the world's longest bridge, opened on January 15, 1927. This vertical lift bridge was constructed under the authority of a 50-year franchise issued to the Dumbarton Bridge Company in 1924. Owned and operated as the Dumbarton Toll Road, it cost its owners $2 million. Joining Menlo Park and Newark, the bridge shortened, by many miles, the distance between the two cities.

Positioned at the center of the moveable span, the toll booth of the two-lane Dumbarton highway bridge caused many a traffic jam. From his booth, the toll collector also pulled the ropes that raised and lowered the warning gates. Photo: CALTRANS.

This 1959 aerial view is of the Dumbarton Bridge, the first highway drawbridge to span San Francisco Bay. Built in 1927, it served for 57 years.
The photo below, taken from the north, shows the lift span of the above bridge at the moment of its demolition. Seen in the foreground is the new Dumbarton bridge and in the distant background, the existing Southern Pacific bridge. Photos: Top, CALTRANS, bottom U.S. Coast Guard.

The 1.2-mile bridge portion of the 6.5 mile structure consisted of 122 steel girder spans and nine 228-foot steel thru truss spans. The lift span was the third from the east end of the bridge. Fabricated in Oakland and barged to the site, the truss spans were lifted onto their piers by the action of the tide. The first paying customer crossed the bridge only 22 months after the start of construction.

Power to lift the span was supplied by a 150 h.p. marine engine. A standby four cylinder diesel engine was used in emergencies.

The toll charge was originally based on the capacity of each particular vehicle and the number of occupants. For example: a two-seater cost 20 cents, plus 5 cents for each occupant and a three to five-seater, 30 cents, plus 5 cents per occupant, etc. At the time the state took over the bridge, the rates stood at 40 cents per car (any size) and 5 cents per occupant. In 1982, the year the bridge was removed, a straight fare of 70 cents per auto was paid by westbound traffic only.

The state bought the bridge in 1951 for $2.5 million making it part of the state highway system. The only major improvement made to the bridge from the time the state took it over until it was removed, was the construction, in 1961, of a new toll plaza on the eastern bridge approach. Previously situated at the center of the lift span, the toll station caused major traffic congestion. Since the warning gates were operated manually by means of ropes from the toll booth, a new electrically operated system had to be installed. Other than the navigation lights, the only lights on the bridge were at each end of the draw span. The 23-foot, two lane roadway was not lighted.

As late as 1971, the bridge raised for water traffic on the average of 120 times a month and was tended 24 hours daily.

In October 1982, a high-level concrete bridge, built to replace the drawbridge, was completed alongside the old structure.

Declared worthless by the highway department, the first and last of the bay's vehicle drawbridges was ordered demolished. At 2:55 p.m. on September 24, 1984, the lift span of the 57-year old bridge was brought down in a spectacular explosion. Later that same week the 186-foot lift towers, and 300-ton counterweights were likewise destroyed. The steel truss spans were all recovered and sold at scrap metal prices. Ironically, the bridge cost nearly a million dollars more to demolish than to build.

Fifteen hundred feet of the approach spans at each end of the bridge were left intact to serve as fishing piers.

Completed two years after the Dumbarton Bridge, the twelve-mile-long San Mateo-Hayward Bridge became the longest bridge in the world. Built at a cost of $6 million, it was sold to the highest bidder in 1968 for $57.20 and then demolished. View is from the southeast. Photo: CALTRANS.

(Hwy. 92) San Mateo-Hayward * Mile 23.5

The Dumbarton Bridge held the record for the world's longest bridge for only two years. With the opening of the San Mateo/Hayward Bridge on March 3, 1929, the title shifted from Menlo Park to San Mateo.

Located approximately 17 miles south of the San Francisco-Oakland Bay Bridge and nine miles north of the Dumbarton Bridge, the $6.5 million San Mateo Bridge measured a full 12 miles, seven of it over water.

Known early on as the San Francisco Bay Toll Bridge, it began as a privately owned venture built under a 50-year franchise issued in 1927 to Messrs Towns, Moon and Duels. Only the second highway bridge to span the bay, President Coolidge, in his last official act pertaining to California, pressed a button in the White House to signal the opening of the "world's longest bridge."

On February 18, 1928, Guano Island off Little Coyote Point in San Mateo was removed by an explosion to make way for necessary bridge pilings. Three months later, 5,000 people gathered to witness the groundbreaking ceremony that signaled the actual start of construction.

At each end of the 303-foot, 613-ton vertical lift span were two 300-foot steel truss spans. Long concrete approach spans reaching out from both sides of the bay made up the rest of the bridge. The steel spans and the precast concrete piers and spans, were built in South San Francisco, Oakland and Alameda, then towed to the site on barges. It is claimed that every ounce of Portland cement used in the construction was made from shells dredged from the area of the bridge.

When first opened, the bridge was marked for three lanes of traffic, but its 27-foot wide roadway proved to be too narrow, and consequently too dangerous. The center "suicide" lane was eventually eliminated leaving only two lanes.

As with the Dumbarton Bridge, tolls varied with the loads. A two-passenger car cost 40 cents; a five-passenger car, 45 cents; seven-passenger, 50 cents. Passengers were charged five cents extra each. Motorcycles were 20 cents (with side cars 30 cents), buses $1.00, and trucks from 60 cents to $1.50.

The State of California paid $6,000,000 for the bridge and took over control on September 12, 1951, making it a link of State Highway 105. The toll was changed to 70 cents for westbound traffic only.

Traffic on the bridge increased from approximately 2,000 cars a day in 1929, to nearly 56,000 by 1968. Bridge openings, averaging ten per day in 1948 also increased during the same 20 years.

All drawbridges and especially older ones, get stuck in the open position from time to time and this bridge was no exception. Mechanical problems, which often caused massive traffic jams, plagued this bridge in its final years. There were also accidents a plenty on the bridge and occasionally an accident to the bridge. One such accident in July of 1967, involving a barge, closed the span for several days.

It was also the scene of a brutal murder in 1933. In about November of that year, two men kidnapped a San Jose man and demanded $40,000 for his return. Before the family had time to pay the ransom, the victim was taken out on the bridge where he was killed. The kidnappers hit him over the head with a brick, tied a weight to him and threw him off the bridge. As he was going over the side, they also shot him several times for good measure.

Not long after the body was recovered, the kidnappers were captured and lynched by a crowd of vigilantes in San Jose. The tree that was used in the hanging was later cut down by the city in an effort to put an end to the episode once and for all.

Early in 1968, the state completed a high-level fixed bridge to replace this aging span. By June of the same year, removal had begun.

After engineers determined the five steel truss sections of the bridge were still usable, the state conducted an international ad campaign to find possible buyers. Murphy Pacific Corporation of Emeryville turned out to be the only bidder. They bought the four truss spans, vertical lift span, lift towers and counter weights, all for a grand total of $57.20. It reportedly cost them another $17,000 to dismantle and tow the sections away. While probably disappointed with the amount they received, the state claimed they still saved the taxpayers many thousands of dollars over what it would have cost to have the dismantling done.

As for Murphy Pacific, it was a case of deja vu. Forty years earlier they towed the very same bridge sections to the site.

In this July 1968 photo, all that remains of the old San Mateo Bridge can be seen in back of the new span. In the process of being dismantled, the sections were hauled away on barges by Murphy Pacific, the same company that brought them there forty years earlier. Photo: CALTRANS.

For the token sum of ten dollars, the County of San Mateo purchased the trestle approach out to where the first truss span had been. The long pier is used by fishermen and others. Under terms of the sale, the county agreed to maintain the trestle and prevent it from becoming a menace to navigation.

The bridge approach on the San Mateo side of the bridge crossed Seal Slough, a natural waterway between Brewer Island and the mainland. Because the waterway was considered navigable, the 43-foot section of the approach over the slough contained a removable section that could be lifted out if necessary. It is not recorded how many times, if ever, the span was removed.

NEWARK SLOUGH

(So. Pacific.) Newark * Mile 0.5

Between the eastern end of the Dumbarton Bridge and the town of Newark, the Southern Pacific tracks cross Newark Slough over a short steel thru truss swing bridge. Built in September of 1908, it was seldom required to open for water traffic. During one 30-year period, it reportedly opened only six times. It also went for 16 years once without a single opening. It has a gas engine for opening the draw, but has no navigation lights. No longer used by the railroad, it is also exempt from opening.

Although built 87 years ago and still in place, this bridge has not been used by Southern Pacific trains for at least a decade. Rarely required to open over the past 25 years, it is now maintained in the open position. Photo: U.S. Coast Guard.

MUD SLOUGH AND COYOTE CREEK

(So. Pacific.) Drawbridge * Mile 0.7 - Mile 4.0

In 1874 the Santa Clara Valley Railroad began construction on bridges over Mud Slough (formerly Warm Springs Creek) and Coyote Creek (formerly Coyote River) on their San Jose-Newark route. Before the bridges were finished in 1875, however, the company became the South Pacific Coast Railroad. Still later, it became the property of Southern Pacific.

The two swing bridges were located one mile apart at the north and south sides of Station Island. The bridgetender's shack, the only building on the island for the first few years, was located equidistant between them. "Scow schooners" made up most of the traffic that navigated Warm Springs Creek to Warm Springs Landing, four miles up stream from the bridge. Both bridges were replaced in 1905 with slightly updated, but similar structures. The Warm Springs bridge suffered some damage the following year during the great San Francisco earthquake of 1906. In 1987 it was replaced with the existing swing bridge.

Expansion of salt ponds in the area reportedly caused the two waterways to silt in to the point of being too shallow for most commercial use. This bridge is no longer required to open for water traffic.

The Coyote Creek Bridge had not opened for water traffic in 15 years when it was finally replaced by a fixed bridge in 1948.

Warm Springs Slough bridge c1910. Located on the north side of Station Island, it was one of a pair of swing bridges erected in 1905 to replace similar ones built by the South Pacific Coast Railroad in 1875. Photo: (Dowd) San Francisco Bay Wildlife Society.

The old Coyote Creek Bridge as it appeared in 1906. Manually operated, it averaged at most three openings a year, but hadn't opened in at least 15 years, when removed in 1948. Photo: U.S. Coast Guard.

Station Island later became know as Drawbridge and eventually was listed in the railroads timetables as such. During the 1880s, two passenger trains stopped at Drawbridge each Sunday. The morning train stopped on the south bound trip to Santa Cruz and the evening train on the trip north from San Jose to Alameda.

In the early to mid-1920s, when Drawbridge realized its height of popularity, it boasted a hotel, 90 residences (mostly seasonal), many gun clubs, a store and a bar or two. Five trains a day made it possible for people to live there and carry on a seminormal life. Children walked or rode bicycles along the railroad tracks to school in Alviso. Some of the more permanent residents took the train to jobs on the mainland.

The bridgetender, according to reports, was often the victim of pranksters. They would purposely have him running from one side of the island to the other to open the bridges for their little skiffs.

Water pollution, lack of fresh water, a general sinking of the island and other environmental problems contributed to the final evacuation of the island. Vandalism also played a part, as many homes were looted and even burned while the owners were off the island.

Today Drawbridge is off limits to the public except for tours conducted by the San Francisco Bay Wildlife Society. The island is under the jurisdiction of the U.S. Fish and Wildlife Service.

COYOTE HILLS SLOUGH

Alvarado * Mile 1.8

Coyote Hills Slough enters the bay via the Coyote Hills Alameda County Flood Control Channel, a dredged channel that drains the Pleasanton Ridge. Much of the area is comprised of salt ponds owned by Cargill (Leslie Salt).

Leslie Salt Company completed a small bascule bridge across Coyote Hills Slough on September 10, 1926. Providing a 39-foot channel over this shallow stream, the bridge reportedly never opened for water traffic during its 22 years of existence. This bridge was removed in 1947-1948.

CHAPTER 4

NORTH SAN FRANCISCO BAY AND TRIBUTARIES

CARQUINEZ STRAIT

Dedicated on November 1, 1930, the $12,000,000, double-track railroad bridge spanning Carquinez Strait represented the final link in Southern Pacific's transcontinental line to Oakland. A massive, mile-long structure, it was the largest and heaviest railroad bridge west of the Mississippi.

The vertical lift bridge connects Army Point in Solano County with Martinez in Contra Costa County. Situated between the bay and the entrances to both the Sacramento and San Joaquin Rivers, it was, for many years, host to more large ships than any other drawbridge in California. However, because it provides up to 71 feet of vertical clearance in the closed position, it opens far fewer times than some other moveable bridges in the state. This bridge is tended 24 hours a day.

Southern Pacific's Martinez Bridge welcomes home a shipload of servicemen from the war in the Pacific. Sixty-five years old in 1995, this huge bridge is still in active service. Photo: CALTRANS

When fully raised, the 328-foot lift span provides 135 feet of vertical clearance and can be lifted to its full height in less than one minute.

This bridge is a far cry from the structure Southern Pacific planned for this location in 1903. At that time, the company requested permission of the War Department to build a wooden trestle drawbridge (type of moveable span not mentioned) across the strait.

The company's announcement of its intention caused an uproar unprecedented in the area. Nearly everyone except the steamboat owners (Southern Pacific) were against the bridge. Representatives of the owners of more than 200,000 acres of delta land opposed it on the grounds that such a structure could cause flooding. Shipping companies claimed a trestle bridge would be a major obstruction on the waterway and would put an end to schooner traffic. The mayors of Benecia, Martinez, Antioch and Stockton were against it as were the directors of 20 plus other cities and districts. An editorial in the San Francisco Examiner stated "Southern Pacific proposed to build a cheap trestle drawbridge across Carquinez Strait." Even the governor allowed that the bridge was probably not in the best interest of the state.

On December 30, 1904, the War Department rejected Southern Pacific's application for a pile trestle drawbridge.

The existing bridge replaced an auto and passenger ferry that had operated between Vallejo and Rodeo for more than 50 years and retired the world's largest railroad ferry, the *Solano*. Until the bridge's completion, the *Solano* carried the company's trains between Port Costa and Benecia.

At one time, Port Costa was reputed to be the greatest grain port in the world. As many as 25 large sailing ships a week were loaded with wheat from the Sacramento and San Joaquin Valleys. In 1884, nearly three million acres of land in California produced upwards of 46 million bushels of wheat.

The phrase "Rats in Port Costa" meant little outside of Port Costa, but to the natives, it described an absolute worse case scenario. Men were known to use it as an expletive when informed their mother-in-law was staying over.

PACHECO SLOUGH

Martinez (So.Pacific.) * Mile 1.1

Pacheco Creek, along with Walnut Creek, Pine Creek and other minor streams, empties into Suisun Bay via Pacheco Slough. These waterways, carrying runoff from the western slopes of the Diablo Range, form one of the major drainageways of Contra Costa County.

A century ago, Pacheco creek joined the tide waters in the tulles near the town of Pacheco. In the middle to late 1860s, Pacheco was head of navigation on the creek and the most important shipping port and business center in the entire county. Schooners drawing up to six feet of water called at its wharves on the high tide. The town, with a population of several hundred, included schools, churches, a few stores, granaries, and several large storage buildings. During 1867 alone, over 700,000 bushels of wheat, plus a large quantity of other products, were shipped from Pacheco.

Unfortunately, the combination of high tides and winter and spring runoffs in the creek, caused flooding that on several occasions washed away wharves and buildings in the waterfront area. Pacheco also suffered at least two disastrous fires and a major earthquake. Each time, the town managed to rebuild. During the flood of 1868, the creek, which normally flowed west of town, changed course and began flowing to the east. Part of the town of Pacheco went up on stilts and a large number of people moved to higher ground. Messrs Pacheco and Galindo, owners of land a short distance to the east, offered free land to

This 1970 photo is of the Southern Pacific swing bridge over the entrance to Pacheco Slough at Avon. Built in 1901, this bridge is nearly identical to the one S.P. built across Cordelia Slough in 1900. Both are still in place and in use.

the flood victims. Many accepted. The area later went on to become what is now the City of Concord.

By 1870, Pacheco Creek was well-established on the east side of town, but silting had made it too shallow for most boats to navigate. Pacheco's days as a river port ended.

In 1901, Southern Pacific built a small swing bridge across the mouth of Pacheco Slough at Avon (Martinez). Built as part of the railroad's Martinez to Tracy line, the 115-foot span, with a single set of tracks and no lights, is manually operated. Twenty-four hours notice is required for opening.

Martinez (Waterfront Road) * Mile 1.0

On November 23, 1916, just a few yards downstream from the Southern Pacific span, on Waterfront Road, Contra Costa County completed a narrow, light-capacity swing span.

Literally worn out by the ever increasing volume and size of the trucks that used the road, it was in extremely bad shape when replaced in 1947. The present plate-girder bridge, looking less like a drawbridge than any other in the state, cost the county $250,000. The only hint from the roadway that this 752-foot structure is moveable, is the pair of warning arms locked to the sides of the approaches. The 170-foot swing span is manually operated and carries no lights.

There is no record to show how many times either of these bridges have opened, but one thing is certain, with 4-feet of clearance at low water, very few boats could pass under them. Because the creek is an important flood control facility, however, U.S. Corps of Engineer dredges have passed through the bridges to clear silt from the drainageway. The occasions reportedly have been few and far between. Twenty-four hours notice is required for opening.

ALHAMBRA CREEK

Martinez (So. Pacific.)

A tidal stream for a mile or more from its mouth, Alhambra Creek has been considered navigable since before the turn of the century.

In 1910, Southern Pacific Railroad built a swing bridge across the creek at the north end of

Berrillessa Street. Just how often the bridge was opened for water traffic is not known, but, with no commercial use of the waterway, it probably wasn't often.

The swing bridge was removed by the railroad in 1924 and replaced with a steel girder fixed bridge. The new bridge, which was said to have been built without permission of the Corps of Engineers, was the source of complaints off and on for many years. Residents claimed that with only about two feet of clearance at high water, the bridge blocked navigation for even the smallest boats and caused flooding. During heavy runoff, especially on extra high tides, debris piled up at the bridge and send water flowing in all directions.

While the swing bridge may not have had more vertical clearance than the present structure, it could be opened to get rid of the debris and let water move more quickly down stream. Through dredging and other means, the problem reportedly no longer exists.

CORDELIA SLOUGH

(So.Pacific.) Suisun * Mile 1.5

For many years, Suisun Slough was the widest and deepest slough in California. By the late 1800s, most of Solano County's products was shipped from the embarcadero at Suisun City. The head of navigation, Suisun City, situated 15 miles above the mouth of the slough, was only 54 miles by water from San Francisco. During the last decade of the 1800s, a brisk trade existed between Suisun and several other ports and landings around the bay.

In 1900, Southern Pacific erected a steel swing bridge at Bahia (formerly Cygnus) across Cordelia Slough, a tributary of Suisun Slough. The 115-foot pony truss swing span (still in place) is equipped with a four cylinder gas engine, but is rarely required to open for water traffic. The bridge provides two 40-foot channels. Still in use on the company's Oakland/ Sacramento line, 24 hours notice is required for opening.

MONTEZUMA SLOUGH

(Sac. Northern) Chipps Island

The Oakland Antioch and Eastern Railroad, later to become the Sacramento Northern, began construction on a drawbridge across Montezuma Slough at Chipps Island in 1912. A massive wooden structure, it had a 236-foot cable-stayed swing span. That bridge served for seventeen years before being replaced by another swing bridge on January 5, 1931.

Because the railroad's substation was located only a short distance from the bridge and, like the bridge, was manually operated, one person acted as both bridgetender and substation manager.

According to one of the last tenders, the railroad stopped using the bridge in 1951. By 1964, the line had been abandoned altogether and the bridge removed. Some of the remaining sections of the old Sacramento Northern route are now part of the Union Pacific system.

In that same year (1912), the railroad also got permission from the War Department to build a bridge across Suisun Bay, from Chipps Island to Bay Point. Because the rail route would be ready long before the bridge could be completed, ferry slips were built on the north and south sides of the bay so the rail cars could be ferried across until the bridge was completed.

Built in 1912 by the Oakland Antioch and Eastern Railroad, This 236-foot cable-stayed swing span was opened and closed by hand. It crossed Montezuma Slough at Chipps Island. Photo: U.S.Coast Guard.

With the bridge over the slough in place and the ferry slips completed, the railroad began train service between Oakland and Sacramento on September 13, 1913. The rail cars were ferried across Suisun Bay via the ferry *Ramon* , a product of nearby Pittsburg.

Construction on the Chipps Island Bridge reportedly began in December of 1913, but it was never completed. The bridge was to have been 1,200 feet long, with two 400-foot thru truss spans and one 140-foot thru truss vertical lift span. Trestle approach spans made up the difference. It was designed to stand 72 feet above MHW (mean high water).

NAPA RIVER

Throughout the early history of Napa County, waterborne travel played an important role in the growth and prosperity of the region. At one time, the Napa River served as a vital link in the transportation of the valley's products. Medium size boats could reach the City of Napa during high tides, and Suscol Landing was reachable on all tides. During low water, larger boats would sometimes have to tie up at Suscol and transfer freight and passengers to stages.

Until dredged by the Corps of Engineers in the 1800s, the river's depth at low tide averaged about five feet for the first twelve miles, then one foot the rest of the way to Napa. After the dredging, the river was maintained to a width of 75 feet and a depth of four feet, all the way to the City of Napa. High tides added about another five feet.

The peak years for river commerce were 1850-1870 when wheat was the main crop of the region. Hundreds of Indians helped in the harvesting and loading of the grain into shallow-draft boats and barges.

The first steamboat to operate on any kind of schedule between Napa and San Francisco was the *Dolphin*, in 1850. Powered by an old locomotive engine, she made the trip down one day and back the next.

In 1907, the Napa Transportation Company was still advertising passenger and freight service on its steamers *St. Helena, Napa City* and *Zinfandel.* Operating daily except Sundays, they ran from the Mission Street Wharf (Pier 2) in San Francisco to Napa, stopping at Vallejo and Mare Island.

During the 1930s, the steamers *Arrow, Napa Valley* and *Frisby* made six daily trips between Napa and Vallejo, even serving meals on board. The Steamer *Gold* out of Petaluma, made weekly trips between Petaluma and Napa.

In the City of Napa during the summer months of the 1850s, the river was so shallow on low tide, citizens easily forded the river. During the winter and high tides, however, that was not an option. In July of 1850, Mr. William Russell received a charter to start the first ferry service in Napa. Just a year later, in 1851, J.B. Howell obtained a franchise to build the first bridge (toll) across the river. He operated the bridge until 1855, when the citizens of the city bought him out for $1,000 and made it a free bridge. A wooden trestle structure, it blocked all water traffic above the center of town.

The barge *Sutter* is shown taking on a load of grapes at Napa in the early 1900s. Photo: California State Archives.

Napa (Third Street) * Mile 19.8

Napa's first drawbridge spanned the river at the foot of Third Street. Built in 1873, the narrow, thru truss swing bridge, with its 105-foot swing span, reportedly cost $9,000. Designed to carry only horse vehicles and foot traffic, the lightly built iron bridge later took on another role. In 1902, The Vallejo, Benecia and Napa Valley Electric Railroad came to the area and, by 1905, received permission to lay tracks on the bridge's roadway. For the next several years, trains shared the bridge with horses, pedestrians and even autos.

While the Third Street Bridge did not open often, the waterway was no stranger to barges of lumber bound for the Hughes Planing Mill on the upriver side of the bridge. It was also popular with youngsters as a fishing bridge. It was removed to make room for the existing fixed bridge that was dedicated at that same site on August 1, 1932.

As highway transportation became more popular and bridges brought places closer together, river traffic declined slowly but steadily. Between 1951 and 1961, tonnage shipped on the napa River dropped from 200,000 to 20,000 tons per year.

In 1968, during construction of the Bay Area Rapid Transit (BART) in the San Francisco Bay area, the Napa River enjoyed a brief surge of business. Sections of the tube that carry the trains under the bay were constructed at the Kaiser Steel Company plant in Napa and towed downriver to the bay.

In 1963, through the joint cooperation and efforts of Napa and Solano Counties, which are divided by the river, a plan was presented to the Corps of Engineers for consideration. The plan called for a deep water ship channel between Mare Island Strait and a turning basin at Asylum Slough, just south of Napa. At 20 to 35 feet deep and 300 to 400 feet wide, "It would allow ocean going vessels to penetrate the Napa Valley and stimulate new growth."

In 1966, the Corps of Engineers suggested an alternate plan that included a realignment of the river. Although the Army's plan would have accomplished the desired results, many felt the plan would spoil the pastoral character of the river. Nothing ever happened either way, but the idea was kept alive for several years.

Napa (Lawrence Street So. Pacific.) * MILE 19.9

On September 21, 1930, many years after the Vallejo, Benecia and Napa Valley Electric Railroad became part of Southern Pacific, it stopped using the Third Street Bridge. The decision was brought about by projected high cost of providing for railroad tracks on the soon to be built new bridge.

Their operation was moved to the S.P. tracks which, by that time, crossed the river via a fixed bridge at Lawrence Street. Built in 1928, the fixed bridge replaced an earlier swing bridge. The construction date of the swing bridge is unknown. Up river from the Third Street Bridge, it seldom had to open.

Napa (Imola Avenue) * Mile 17.6

For the first time in at least 17 years, Napa once again had a drawbridge. Built by Erikson, Phillips and Weisberg of Oakland, the Maxwell Bridge, named for Thomas Maxwell, was dedicated on October 1, 1949. Located on Imola Avenue, it is part of State route 29.

Four hundred and fourteen feet in length, with a 132-foot lift span, this vertical lift bridge provides up to 60-feet of vertical clearance. The bridge seldom has to open, however, because most boats that far up river can pass under the closed bridge.

The Maxwell Bridge as seen from the south. With adequate clearance for most vessels, this span does not open often.

During the excavation for the bridge approaches, the contractors uncovered a two-hole privy. Being more than 50 years old, it automatically qualified for historical landmark status. Federal law requires archeological research on such landmarks which, in this case, delayed the bridge project and added $37,000 to the cost. It is estimated that research such as that adds about 1% to highway construction costs.

Seventy-two hours notice is required for opening this bridge.

(So.Pacific.) Brazos * Mile 10.6

In 1888, the Santa Rosa and Carquinez Railroad, a subsidiary of the Southern Pacific Railroad, built the first railroad bridge across the Napa River. Located approximately seven miles south of Napa, it is believed to have been a swing bridge.

Southern Pacific replaced that structure in 1903 with a steel thru truss swing bridge. Although located at Dutton's Landing, the bridge was widely known as the Brazos Drawbridge.

On the company's Napa Junction to Shellville line, all rail traffic serving northwestern California passed over that bridge. During the 1930s the bridge was opening for water traffic as many as 2,000 times a year. In 1938, the company started maintaining the bridge in the open position, closing it only to let trains cross.

Due to heavy fogs in the area, special conditions prevail at this location. When the draw is closed for the passage of trains and the visibility at the bridge is less than one mile, the bridgetender must sound two prolonged blasts every minute. When the draw is reopened, the tender then sounds three short blasts.

Until the late 1960s, the bridge had no electricity. A model-A Ford engine operated the drawspan and kerosene lamps served as navigation lights.

Due to its narrow draw (80-feet), the swing bridge had more than its share of mishaps. On June 17, 1947, for example, the 450 foot Union Oil tanker *La Placentia* ran aground while being towed to Napa for scrapping. Not quite clear of the draw at the time of the grounding, it caused damage to the bridge and tied up train traffic for many hours.

A much more serious accident happened on July 12, 1959. Three Southern Pacific locomotives, each weighing 185-tons and valued at $600,000, plunged into the river at the bridge site. A caboose followed them in, but only partly submerged. One of several versions of the incident claimed the signals got crossed letting the locomotives move onto the bridge while in the open position. By the time

In this 1968 photo, a 324-foot section of the twin-bore BART tube is being towed through the draw at Brazos. With only 75 feet of horizontal clearance, it was a challenging feat under the best of conditions. Three Harbor Tug & Barge Company tugs are doing the job. Photo: U.S. Coast Guard.

the engineer saw the open bridge it was too late to stop the train. Another version claimed the brakes failed to stop the train. The three-man train crew suffered only slight injuries.

The Ben C. Gerwick Construction Company of San Francisco raised the locomotives from the river bed with the aid of a 75-ton derrick barge. Submerged for 17 days, there was no guarantee they would be salvageable, but the railroad had no choice but to retrieve them. As it turned out, they were later declared useless and sold for scrap.

Gerwick, on the other hand, came out much better on the deal. Because of the urgency of the situation, engines in the water and the bridge out of service, etc., Southern Pacific insisted the contractor more or less name its own price for the job. The normal procedure would have been to bid the job on a cost plus basis. However, since it was sight unseen, the contractor bid $90,000, an amount he considered outrageous. Southern Pacific accepted it.

According to Mr. Gerwick, they made a small fortune on the job. Coming, as it did, in a year that was not all that profitable otherwise, it didn't go unnoticed. At the company's next annual board meeting, the CEO jokingly announced that, from then on, the company was going to concentrate on finding and raising wayward locomotives from the deep.

In 1979, after 75 hard years of service, the old Brazos Drawbridge was replaced by a vertical lift bridge. Opened to rail traffic on May 17, 1979, the new bridge has a 228-foot lift span that provides a horizontal clearance of 157 feet. Erected adjacent to the old bridge, it too is maintained in the open position, closing only to let trains cross. Its maximum vertical clearance is 97 feet. The bridge is situated on a sharp bend in the river south of Green Island Road.

The new Southern Pacific Brazos Bridge, with 157 feet of horizontal clearance, stands ready to welcome the largest vessels able to reach that point on the river. Four years after its completion in 1983, two locomotives "got away," rolled off the end of the opened bridge and plunged into the river.

The 24-hour changeover from one bridge to the other began on May 16 when, on the afternoon high tide, the swing span was lifted off its center pier and towed away. The center pier was blown to bits later that night. As the swing span left the area on barges downriver, the new lift span, also on barges, started its trip up the Napa River to the bridge site. The approach spans and lift towers were built in place. Installation of the new lift span began on the morning of the 17th. The channel was closed to navigation until May 23.

On October 28, 1983, the new bridge experienced an incident that was remarkably similar to what happened to the old swing bridge in 1959. According to a company spokesman, "The train crew left two locomotives on the main line while they moved another engine to a siding. While they were away, the 'waiting' engines somehow got away." With the span in the raised position, the engines rolled off the end and into the river. Only the lead engine sunk out of sight and, being unmanned, there were no injuries.

DUTCHMAN SLOUGH

James Irvine

A short tidal stream extending northwestwardly to South Slough, Dutchman Slough enters the Napa River just above the Vallejo Bridge. In 1934, James L. Irvine completed a private drawbridge across Dutchman Slough between Island Number One and Knight Island.

A wooden bascule span with timber approaches, it carried an 18-foot roadway and provided a 20-foot channel. Ten feet of vertical clearance on a high tide and 72 hours notice required for opening was almost a guarantee that it didn't open often. Manually operated, it could be raised with the aid of a tractor or a come-a-long fastened to the lift cables.

In June of 1977, Leslie Salt Company, owners of the property replaced the drawbridge with a removable span bridge.

Vallejo * Mile 3.5

On December 19, 1927, the Sears Point Toll Road Company dedicated a bridge across the Napa River at Vallejo. A long wooden trestle causeway with a bascule span over the shipping channel, it carried two narrow traffic lanes. When the state bought the toll road in 1938 it also became the new owner and operator of this and the road's two other drawbridges.

On May 6, 1942, the California Division of Highways applied for permission to build a vertical lift bridge a few feet upriver and parallel to the then existing bascule bridge. The new bridge, also with a 75-foot clear channel, would have become part of the state highway system as Solano Route 208-A. The plan was to have the new bridge carry west-bound traffic and the old bascule bridge, the east-bound traffic.

With the country at war, traffic to and from the Mare Island shipyard was already near the critical point. Nearly 12,000 vehicles crossed the bridge on an average working day. In a letter to the U.S. Corps of Engineers, the urgency of the need for an additional bridge was pointed out and immediate attention to the application was requested. Due to the shortage of steel for anything but the war effort, however, <u>that</u> bridge was never built.

Unlike the other two bridges on the toll road, this was a very active span. In 1941, for example, 1,418 vessels passed through the draw going up stream and 1,420 down. During its last five years of service, it was still averaging 90 openings per month. Most of the water traffic was comprised of pleasure boats, small freighters and towed barges.

The narrow navigation channel, strong tides and occasional dense fogs made passage through the draw an adventure. Those factors accounted for most of the 45 major ship /bridge collisions during the 14-year period 1940 to 1954.

One rather spectacular accident occurred on April Fool's day 1964. A lone tug, towing a 3,500-ton load of salt downstream, tried to pass through the draw. What would have been a difficult task for two large

This aerial view taken on June 19, 1963, shows the progress of the new high-level bridge being built to replace the old drawbridge at right. Rows of Navy ships are lined up at Mare Island. The Navy installation is slated for closing in a few years. Photo: CALTRANS.

tugs, proved to be impossible for this small tugboat and its captain. A strong northwest wind caught the aft end of the barge, causing the forward end to strike the pilings and timbers under the bridge. The force of the blow tore away most of the west entrance to the bridge. The span's operating mechanism was unharmed, but the resting plate, the heavy steel ledge that supported the moveable span in the closed position, was knocked out of kilter. The highway department engineers and their crews worked furiously to shore up the west end of the bridge and realign the resting plate. Occurring at the start of the evening rush hour from Mare Island, the accident caused a monumental traffic jam. Due to the efforts of the highway crews, traffic was flowing again by 7:00 p.m.

On August 10, 1967 a high-level, concrete fixed bridge was completed a few yards north of this bridge. The state removed the draw span and the west approach span soon afterwards. The wooden approach span on the Vallejo side was left in place and today is a popular public fishing pier.

Mare Island * Mile 2.8

In 1919, the U.S. Navy built a long causeway across the Napa River about a mile south of where the Vallejo bridge was later erected. Near the center of the river, over the navigation channel, was a small bascule drawspan. Because of the presence of the Mare Island Naval Shipyard, this section of the river is known as Mare Island Strait.

Although the island became a government installation in 1852, for some reason, it took 67 years to get a bridge to the mainland. During those years, all shipyard personnel and supplies had to be transported to and from the island by boat.

That original bridge consisted of two side-by-side timber trestle approaches that met at the bascule span. One trestle carried autos and foot traffic, the other, railroad tracks.

Ferry operators feared the bridge would put them out of business, but thanks to the many problems that plagued the structure, mechanical and otherwise, their business continued to thrive.

The moveable span and trestle approaches suffered damage from many a wayward craft and accidents were responsible for most of the major repairs over the years. With only 75-feet of horizontal clearance, its

This rare photo shows both the 1919 Mare Island Bridge (background) and its 1935 replacement (foreground). The view is facing the mouth of the strait, with Mare Island on the right. Photo: Vallejo Naval and Historical Museum.

A large Navy ship, under tow, approaches the draw of the Mare island bascule bridge heading down stream. With only 75 feet of horizontal clearance, it was often a tight fit. Completed in 1935, this bridge was replaced in 1980. Photo: U.S. Coast Guard.

approaches and fenders were hit often. A submarine once missed the draw completely and got stuck in the pilings of the wooden causeway.

In 1933, construction began on a new 4,500-foot long bridge up river from the old span. Costing the Navy $901,900, it was built by Barrett and Hilp of San Francisco.

The new bridge carried autos, pedestrians and trains on the same 30-foot roadway. Built of reinforced concrete slab spans on concrete pilings, the new bridge reached out from each side to meet the new bascule span. Because the navigation channel was closer to the island at the new location, possibly due to dredging, the bascule span was also closer. To the disappointment of many, the draw of the new bridge had not been widened, but left at 75 feet. What had already been a problem on the earlier span, was going to be a major one on the new bridge.

Mare Island Shipyard, at the peak of its wartime production in 1945, built 392 ships. That included 19 submarines, five submarine tenders and 31 destroyer escorts. It also built a few hundred landing craft of various kinds. As the Navy ships increased in size over the years, it became more and more difficult for them to fit through the bridge's narrow draw. It was not unusual for the largest ships to have to adjust their ballast to make them list away from the open leaf.

The installation of a new steel deck on the draw span in 1960 was, reportedly, the only major repair to the bridge during its 45 years of service. Its concrete construction and better lighting are credited for the reduction in the number of accidents and amount of damage to the structure.

By 1972, the Navy had already considered replacing this bridge. In 1977, a delegation of Solano and Napa County government leaders and industrialists appeared before the Military Construction Committee in Washington D.C. to present their case for a new bridge. They argued that the 75-foot draw of the existing bridge was too narrow for many ships and especially the 100-foot wide oceangoing barges. They claimed it was costing them all financially. A clearance of 140 feet would end the bottleneck and could generate as many as 1,000 jobs in the area above the bridge.

The Navy was in favor of the new bridge, as were many major companies, local governments and individuals. The project was passed by Congress and work was started in early 1979.

By March of 1980, the towers of the vertical lift bridge were in place and the old bascule span removed. The main and rest piers were destroyed by explosive charges. The 140-foot, 682-ton lift span was fabricated in the Kaiser yard in Vallejo and brought to the site on two barges. On March 29, after very careful maneuvering, the span was brought into perfect alignment and attached to cables from the towers. Once the weight of the span was supported by the towers, the barges were ballasted down until free.

The existing Mare Island Bridge as seen from the Vallejo shore. The second of three vertical lift bridges to eventually span the Napa River, its fate is now in limbo. With Mare Island Naval Base scheduled to close by the end of the century, A drawbridge to the island may become unnecessary.

With the weight of the bridge's counterweights providing a perfect balance, the two 60 h.p. motors easily lift the huge span to its full 100-foot height in 19 minutes. The deck of the bridge, ten feet above the water, allows many boats to pass under it. Even some larger craft require the span to open only part way. The operator's shack is in the west tower, over the roadway.

The new drawbridge was first opened to water traffic on April 7, 1980 and to vehicles on April 29. The ribbon cutting and official dedication of the $13.5 million bridge took place on May 4. Ernest D. Wichels, a Mare Island employee, was on hand for the dedication. He was also on hand for the dedications of the 1919 and 1935 bridges. The same scissors used at the 1935 dedication were used for this ceremony.

The new bridge has three traffic lanes and a set of railroad tracks. Due to past accidents, a double set of lighted barriers and traffic signals are at each end of the moveable span. The old bridge had only a warning light and single set of barriers at each end. There were three fatalities on the bridge during its 45 year history.

During the nearly two months required for the changeover between bridges, auto and pedestrian traffic was detoured via north gate. Railroad service was suspended.

Although tended from 6:30 a.m. until 10:00 p.m., the bridge is exempt from opening during morning and afternoon rush hours. Two hours notice is required for opening at other times.

All upriver ship traffic passes through this draw, making it the busiest on the river. With 100-foot draws in this and the Brazos bridge, large ships and barges are now free to navigate the river farther than ever before. Unfortunately, some of the upriver industries, after enduring years of inconvenience, had already moved downstream by the time the new bridge was a reality.

SEARS POINT TOLL ROAD

On July 4, 1928, the Sears Point Toll Road opened to the public and became an immediate and overwhelming success. Built at a cost of $900,000, it included drawbridges over the Napa River in Vallejo; Sonoma Creek, at the Sonoma-Solano County line; and Tolay Creek at Sears Point.

Ten and 1/4-miles long, the road skirted the shore of San Pablo Bay, from Sears Point to Vallejo, much as it does today. It saved motorists approximately 22 miles, compared to the Vallejo-Napa highway. The toll road provided motorists with a direct route between State Highways 40 and 101. The Vallejo-Napa highway route featured eight miles of gravel road, eight railroad crossings, one city, several small towns and numerous curves and long grades.

In 1925, A. G. Stewart, president of Sears Point Toll Road Company, a company wholly owned by Golden Gate Ferries Inc., obtained a fifty year franchise to build and maintain the road and collect tolls. The idea behind the venture was to promote development in the areas served by the company's near monopoly ferry business.

The company paid the state $100 per month license tax and charged tolls at the following rates: one cent per hog or sheep; five cents per person on foot or auto passenger; twenty-five cents for autos; thirty-five cents for trucks; and ten cents each for horse-drawn vehicles and motorcycles.

Nearly every county government and chamber of commerce in the entire bay area was 100% in favor of constructing the road. Only the City of Napa, which stood to lose business by the route change, objected. Some industries on the Napa River were unhappy with the prospect of another drawbridge for their boats to go through, but otherwise favored the idea.

Whether the road was as successful financially as it was popular with the motoring public is not certain. In 1938 the toll road was sold to the State and became part of State Route 37.

In 1964, Senator Luther Gibson submitted an amendment to the State Legislature to study the idea of making the route, still called Sears Point Road, a toll road again in exchange for giving it freeway status. According to his plan, the three drawbridges would be replaced by high-level fixed bridges.

His amendment never passed, but at least part of his idea came to pass, now that all of the drawbridges have been replaced by high-level spans. The good-news/bad-news is that, while the road remains free, it still is not a freeway.

SONOMA CREEK

Sears Point Road * Mile .1

Sonoma Creek flows southeasterly from the hills above St. Helena and enters San Francisco Bay just below its junction with Napa Slough. It has been classified as navigable since 1886, for a distance of approximately 10 miles above its mouth.

During the 1850s and 1860s, tugs brought barges and schooners to Norfolk Landing where they loaded agricultural products and meat. A large slaughterhouse operated near what is now Schellville.

In October of 1850, regular tri-weekly steamer service was inaugurated between Sonoma embarcadero and San Francisco. The steamer *Georgia* made the first passenger run on October 11, 1850. From Sonoma, passengers could continue by stage to Napa, Benecia or even Sacramento.

The first and only drawbridge to span Sonoma Creek along the Sears Point Toll Road stood astride the Sonoma-Solano County line. Completed on December 17, 1927, it consisted of 1,000 feet of wooden trestle approaches and an 80-foot steel, pony truss bascule span. With 17-1/2 feet of vertical clearance at low water, nearly all water traffic could pass under the closed bridge.

By the 1960s, Sonoma Creek was no longer a busy waterway. An occasional sand barge towed from Napa Slough or a "lost" fishing boat, made up most of the traffic through the bridge.

Built in 1921 by the Northwestern Pacific Railroad, this Scherzer rolling lift bridge at Wingo now belongs to the California Northern Railroad. Although it offers practically no vertical clearance at high tide, it opens only once or twice a year for water traffic. Photo: U.S. Coast Guard.

In 1967, spite of the inactivity of the drawbridge (average of one opening a year) there were many objections when the state revealed plans to replace it with a fixed bridge with 21 feet of vertical clearance. The state compromised and the bridge was built with 23 feet of clearance.

Erected down stream from the drawbridge, the new reinforced concrete span was completed in 1969. The old bridge was removed soon afterwards.

(N.W.P.) Wingo * Mile 5.4

The earliest railroad bridge to span Sonoma Creek was a swing bridge built near Wingo in 1879 by the Sonoma Valley Railroad. That bridge was replaced in 1921 by the existing bascule bridge, a Scherzer Rolling Lift type. Built by the Northwestern Pacific Railroad, it opened to rail traffic on July 20, 1921.

Located just up river from the old bridge, it provides 44 feet of clearance between fenders. Seventy-three years old and averaging less than two openings a year, it is still in use by its new owners, the California Northern Railroad.

(Sonoma Land Company) Wingo

In 1904, the Sonoma Land Company (A.G. Skaggs) built a bascule bridge across Second Napa Slough near Wingo. A wooden trestle bridge, with a small bascule span, it stood until 1921, when it was replaced by a second bascule bridge. Without having been moved an inch, this bridge has been listed as spanning Sonoma Creek, Steamboat Slough and Napa Slough. Second Napa Slough joins Sonoma Creek.

When the Navy took over much of the surrounding area many years ago, it also became the owner of the bridge. The drawbridge was eventually replaced by a concrete structure.

STEAMBOAT SLOUGH

Wingo-Schellville

The Sonoma Valley Railroad also built a drawbridge over Steamboat slough, near Schell (now Schellville) in 1879. On September 15, 1906, the company was ordered to repair that bridge or remove it within 60 days. It is not known for sure which option the company took, but it was probably the former since a swing bridge at that location burned to the waterline in c1948.

TOLAY CREEK

Sears Point Road * Mile 3.7

A few yards from the western end of the toll road, where it joined the San Rafael-Napa road, the Sears Point Toll Road crossed Tolay Creek. A short, shallow tidal stream, formerly Midshipman Slough, it connects with Sonoma Creek via a connecting slough.

In 1915, Tubbs Island Company built a wooden, hand-operated bascule bridge across the creek to land it was reclaiming for agriculture. In 1926, the Sears Point Toll Road Company built a similar bascule bridge 75 feet downstream from the Tubbs bridge. As per their agreement with the S.P.T.R.Co., the Tubbs Island Company removed their bridge in exchange for free passage over the new one. The Tubbs Island Bridge reportedly stuck in the open position during its initial inspection in 1915 and had to be forced closed. It eventually passed inspection, but may never have had to open again for boats or any other reason.

Two years after the first Tolay Creek Bridge was completed, and shortly before the road opened to the public, the new bridge burned to the ground.

It was replaced on June 12, 1928, by a steel vertical lift bridge. The 50-foot moveable span was manually operated by a crank.

Above is the first Tolay Creek Bridge built by the Sears Point Toll Road Company. Completed in 1926, this manually operated wooden bascule bridge burned to the waterline after only two years. Photo: U.S. Coast Guard.

Shown in the photo to the left is the steel, vertical lift bridge (still hand operated) built in 1928 to replaced the above structure. The view is to the east and shows Hwy. 121 intersecting with Hwy. 37. Photo: CALTRANS.

The narrow, two-lane bridge was not required to open often, but it did average one or two openings a year for awhile. Barges hauling rocks from upstream accounted for what little traffic there was.

By 1949, the bridge was so severely neglected, (bolts and cables rusted through, crank frozen and counterweights secured to the towers to keep them from falling onto the roadway), the state removed the towers and weights. After that surgery, all that remained were the wooden approaches and the immovable pony truss lift span. On October 4, 1952, the structure was declared a fixed bridge and twenty years later, in 1972, a concrete bridge took its place. Smothered in tulles, Tolay Creek has all but disappeared.

(N.W.P.) Grandview * Mile .8

In 1888, the Northwestern Pacific Railroad built a swing bridge over the Petaluma River just above its mouth at Grandville. Situated on its Schellville branch line, the bridge served until 1911, when it was replaced by the existing bridge.

Opened to rail traffic on November 26, 1911, the new bridge's 250-foot swing span is powered by a gas engine. Records for 1960 indicate that it was tended 24 hours a day and opened 2200 times during that particular year. As with most railroad bridges today, it is kept in the open position and closes only for trains.

Above: The original Northwestern Pacific Railroad bridge at Grandville as it appeared in 1891. Photo: Marin County Historical Society.
Below: The existing NWP bridge, built in 1911. The town of Black Point is in the right background.

In the early years, The railroad kept a tug at the bridge to aid sailing ships and scows through the draw.

Black Point * Mile 1.0

On June 14, 1917, a large and important highway drawbridge was dedicated at Black Point (formerly Green Point). Spanning the Petaluma River at the Marin-Sonoma County line, it shortened, by many miles, the distance between most Marin County cities and points in Napa and Solano Counties. Prior to the bridge, it was necessary to go through Petaluma, to get to Sonoma, Napa or Vallejo.

A typical Sunday automobile trip in 1917 might have gone something like this: from San Francisco, south to San Jose then north on the east side of the bay to Oakland, Richmond and Martinez. From Martinez to Benecia by ferry, then to Vallejo, Napa, Black Point and Sausalito. From Sausalito it was the ferry back to San Francisco. With a total distance of about 260 miles, one can be sure there were no lengthy stops during this all-day trip.

The $10,000 Strauss bascule bridge, one of the largest of its type in the state, weighed 625 tons. Perfectly balanced, its 150-foot leaf could be moved with one hand when its gears were disengaged. In actual use, however, the huge leaf was raised and lowered by two 25 h.p. electric motors. During the years leading up to the actual construction of the bridge, there were the usual arguments for and against the planned structure. One letter to the Corps of Engineers came from the owner of 12 scow schooners that plied the river several times a week between San Francisco and Petaluma. He wrote that he had no objections to the bridge itself, but insisted the state should provide a tug to help his sailing scows through the draw like at the railroad bridge down river. Unfortunately for him, the government felt otherwise. The Corps of Engineer's stated they felt Northwestern Pacific

The underside of the Black Point Bridge viewed from the eastern approach. The perfectly balanced 150-foot leaf was raised and lowered by two 25 h.p. electric motors. This important bridge opened to traffic on June 14, 1917. After serving only 41 years, it was replaced by the existing concrete structure. Photo: California State Library.

Railroad's reason for providing a tug was probably as much to protect the bridge as to help the ships. If the owners of the scows chose, they could station a tug at the bridge, but it was not up to the state.

As with all drawbridges, this one no doubt had its share of minor accidents on and under the span, but not much has been written about them. However, in what appeared to be a last gesture to be remembered by, the old bridge, with the help of an inattentive operator, knocked a pile driver into the river in 1957. The incident occurred during the construction of the bridge that was being built to replaced it.

The west side of the Black Point bascule bridge showing the huge counterweight and operator's house. Photo: Shirley Ghiselli Coleman.

For some months, tugs had been pulling barges loaded with pre-cast piles, girders and deck sections down river from the Ben C. Gerwick plant in Petaluma. Each time the tug and its tow passed through the bascule span, the operator had to lift the leaf only about one quarter of the way to let it through. Then came the day when the contractor needed a pile driver brought down to the site. Being an old model pile driver, it stood tall, too tall, unfortunately. The bridgetender recognized the tug, but failed to see the huge piece of equipment on the barge. Apparently fooled by his long association of tug and tow, he became complacent. The tug made it through, but the pile driver hit the leaf and went into the river. Damage to the huge bridge was not serious.

The existing concrete high-level bridge was completed in 1958, without further interference from the old bridge. As a side note, Ben C. Gerwick Sr., one of the major contractors on the new span was the state's resident engineer during the construction of the old bridge.

PETALUMA

Petaluma has been the head of navigation on the Petaluma River since the beginning of its use for commerce. Until 1961, however, when the 86th Congress changed its name, this waterway was officially Petaluma Creek. Not nearly as well known as the Sacramento or San Joaquin Rivers, it was at one time California's second most important river in terms of tonnage carried. In 1911, Petaluma was also the state's third most important inland port.

Under the jurisdiction of the U. S. Corps of Engineers since 1880, the river is maintained to approximately 100-feet wide by 3-feet deep at low water. In 1896-97, a 3,000-foot canal was cut from the river. Although privately funded, the use of the canal and turning basin is free to all. Several warehouses and mills were built along its banks.

In the late 1850s, freight and passenger vessels were plentiful on the "Creek," as it was called. Shipping in 1896 amounted to nearly $11.5 million and included wheat, wood, eggs, silk and general merchandise. Ferries between San Francisco and Petaluma operated for nearly 100 years (1852 to 1950). During the later years, over 12,000 passengers a year traveled via the Petaluma Ferries.

Each week, thousands of dozens of eggs left Petaluma for the San Francisco markets on strange looking scows affectionately called "Egg Boats."

Petaluma Landing c1915. The schooner *Margaret C*, can be seen tied to the dock ahead of the *Crockett* and *Montezuma*. The probable location is between "D" and Washington Streets. Photo: San Francisco Maritime Museum.
Below: The Haystack Landing Bridge as seen from Highway 101. This bridge is maintained in the open position.

Haystack Landing (N.W.P.) * Mile 12.4

The San Francisco and North Pacific Railroad built an iron swing bridge across Petaluma Creek at Haystack Landing in 1876. After years of problems having to do with the span not opening fully, and numerous charges of it being a menace to navigation, the company agreed to replace the ailing bridge.

Plans for the new structure were approved on November 21, 1903 and the bridge was completed sometime in 1904. Originally manually operated, it was electrified in October of 1906, becoming the first mechanically operated bridge in the north bay.

With only one draw (west side) and a skimpy 68 feet of horizontal clearance, it has barely enough room for some of the barges that pass through. In the 1960s, it averaged nearly 1500 openings a year for water traffic and was tended 24 hours a day. Today it is maintained in the open position and rail service over the line has ended. This was the docking area for the ferries and egg boats during their heyday on the river.

"D" Street * Mile 13.7

The first drawbridge at "D" Street was built in 1883. A manually operated iron and wood swing bridge, it suffered from a faulty mechanism and was hit often by passing boats. On more than one occasion, tugs or other watercraft were called upon to aid in the opening or closing of the draw span. By 1920 it was declared obsolete and plans were being made for a new bridge.

It wasn't until July 5, 1933, however, that the City of Petaluma dedicated a new bascule bridge at "D" Street to replace the 50-year old swing bridge. A rivetted, pony truss span with a single 80-foot leaf,it supports a 32-foot roadway and two sidewalks. With its counterweights and machinery below the street level, it is less conspicuous than most bascule bridges.

The existing "D" Street Bridge. Built in 1933, it replaced a swing bridge completed in 1883. The view is from upriver.

A busy bridge, it opens approximately 1400 times a year, primarily for pleasure and fishing boats. The number would be much greater except that its vertical clearance of 13 feet at low water allows many boats to pass under the closed span. Although four hours notice is required for daytime openings, a bridgetender is often at the bridge all day on a busy summer weekend day. Twenty-four hours notice is required between 6 p.m. and 6 a.m.

Petaluma is a destination for many yachting clubs, partly because there is a fine turning basin and mooring facilities in the heart of the downtown business district. The trip on the river, with its winding pastoral setting, is another reason for the area's popularity.

The ferries and the "Egg Boats" are long gone, but, after nearly 150 years, the river is still active and Petaluma is still the head of navigation. Since the summer of 1993, the *Petaluma Queen,* a sternwheeler, operating out of the turning basin, conducts weekend dinner and luncheon "cruises" on the river. At the start and finish of each trip, the Queen passes through the "D" Street Bridge.

Washington Street * Mile 14.2

One of the earliest drawbridges in the state was completed at Petaluma in May of 1857. A small wood and iron swing bridge, it was erected by a real estate company to get potential customers over to the then mostly uninhabited east side of the river. Located at Washington Street, it was washed out a year or two later by one of the frequent floods that occurred each spring.

It is not known just when that bridge was replaced, but on December 9, 1864, a wooden swing bridge was built on the site to replace an existing bridge. Built at a cost of $823, this bridge served until 1914, when the city replaced it with a modern looking, but still manually operated, drawbridge.

Although Petaluma probably could have gotten along just fine without a moveable bridge at this location, it was built anyway and survived for 56 years. Photo: CALTRANS.

Dedicated on August 29, 1914, the new double-leaf bascule bridge carried a 24-foot roadway and two 5-foot sidewalks. Its two 30-foot leaves afforded more than adequate horizontal clearance for boats to pass through the bridge.

Building this bridge, at a cost to the taxpayers of approximately $16,000, did not happen without a struggle. Fourteen-plus miles from the mouth of the river, Washington Street was the highest point reached by nearly all vessels. Only small craft could, or had any occasion to go beyond it. Many people objected to "building a moveable bridge where it wasn't needed." One resident even found a 1901 news article from the local paper that claimed the creek was no longer navigable beyond the old swing bridge, which should be removed.

The more optimistic civic leaders, those who refused to believe the river had seen its best years, won out. It may not have opened often for water traffic, but it served a useful purpose and was part of the Petaluma scene until April of 1970, when it was replaced by the existing concrete bridge.

NOVATO CREEK

Novato Creek flows southeasterly, entering San Pablo Bay just below the mouth of the Petaluma River. This tidal stream, considered navigable since 1880, has been host to at least four drawbridges. It is not known how many times any of them had to open for water traffic, but it was probably not many.

Before the turn of the century, boats of up to 25 tons made weekly trips between Novato to San Francisco. Novato, like Petaluma, was an egg producing region, but it also shipped a variety of agricultural products to the San Francisco market. It looked for awhile like Novato Creek had the potential of becoming an important waterway. Unfortunately, it never happened. The coming of the railroad to the area in 1878 is blamed for putting an end to whatever promise it might have had.

Bel Marin Keys * Mile 0.9

A 1,670 acre water oriented residential community, Bel Marin Keys is situated below the high water level of Novato Creek where it is protected by a system of dikes.

Situated due north of the old Hamilton Field Air Base, the area was diked and drained many years ago to reclaim it for agriculture, mostly hay. After being developed by Bel Marin Keys Development Corporation, the area now consists of several lagoons connected by canals dredged to approximately 20 feet. Boats enter and leave the lagoon through navigation locks. A secondary function of the locks is to provide an inlet from Novato Creek to regulate the height and quality of the water in the lagoons.

This private, retractable drawbridge spans a navigation lock at the end of Bel Marin Keys Boulevard. Connected to Novato Creek, Bel Marin Keys is a large waterfront home development near Ignacio. Photo: U.S. Coast Guard.

In 1985, a retractable drawbridge was built across the lock situated at the end of Bel Marin Keys Boulevard. The drawbridge makes it possible for boats to enter the 60-foot lock unimpeded while still allowing farm and emergency vehicles to cross the lock to reach the levee road beyond.

Riding on two sets of 12 inch wheels, the 13.5-foot by 45.8-foot roadway rolls horizontally across the 18.6-foot width of the lock. The bridge section is extended and retracted by means of an underhung cable sheave-and-winch system powered by an electric motor.

The bridge is maintained in the open position and is closed only occasionally for the passage of vehicles. Locked gates at each side prevent vehicles from accidentally going into the water. Vertical clearances under the bridge when closed are 9.8 feet at MLLW and 4.5 feet at MHW.

(S.F.&N.P.) Novato * Mile 5.3

The first drawbridge across Novato Creek was built by the San Francisco & North Pacific Railroad on its San Rafael to Petaluma line. Constructed in 1888, the manually operated structure was located about five miles above the mouth of the creek.

In March of 1917 a new bridge was built at that location. Although reportedly still in place during the late 1930s, it was classified as a removable span bridge by 1925.

Thomas Keena, a resident of Novato since 1912, remembers the railroad drawbridge spanning the creek about 1/2 mile south of the Novato train depot. "In 1921, I used to swim under it at high tide," he said. "The pilings where the boats used to tie up were still there, but little was left of the creek except a deep area under the bridge." Mr. Trumbull, manager of the Novato Land Company, told Mr. Keena that the bridge opened for the last time in c1910-1912, for a barge load of horse manure from San Francisco's Fisherman's Wharf. The manure was bound for Sweetzer Landing, the original landing where Novato had its beginning. Mr. Trumbull remembered the occasion because he had to phone the railroad to have the bridge opened that last time.

(S.F.& N.P.) Ignacio * Mile 3.7

In c1878, the same railroad constructed a narrow gauge line from what is now Ignacio, to Sears Point. In c1880, the railroad erected a small manually operated bascule bridge over Novato Creek just north and east of Ignacio, about 3 1/2 miles above the mouth. It was replaced by a sturdier bridge in 1890, when the line, which operated as the Sonoma Valley Railroad, went from narrow to standard gauge.

Once again, on March 31, 1917, the Northwestern Pacific Railroad replaced that bridge with yet another bascule span. Because of a shift in the creek's channel, however, the bridge was built a few feet to one side of the then existing span. It was removed sometime after 1949.

(Hwy. 37) Ignacio Mile * 3.7

In March of 1918, a wooden highway bascule bridge was erected over the creek on Hwy. 37, about 3/4 of a mile east of Hwy. 101 and parallel to the Sonoma Valley Railroad bridge. That manually operated structure survived until at least 1929 before being removed or converted to a removable span bridge.

The Hwy. 37 bascule bridge over Novato Creek in 1928. The view is looking south. Photo: California State Archives.

Highway 101 * Mile 3.2

When the Richardson Bay Redwood Bridge opened on Sunday November 22, 1931, it brought an end to a nightmare traffic situation that had bottled up southern Marin County for years. Until the bridge was built across this shallow arm of San Francisco Bay, all north and south vehicle traffic, between the Sausalito ferry depot and San Rafael, followed a circuitous route around the western edge of Richardson Bay. The new route, essentially from Waldo to Alto, shortened the distance and eliminated considerable backtracking and traffic congestion. The bridge spanned the bay from Manzanita to De Silva Island and included an over-crossing of the Northwestern Pacific Railroad tracks.

The Richardson Bay Bridge "Redwood Bridge" c1950. The view is to the northwest. Photo: U.S.Coast Guard. Bottom: November 22, 1931, Opening Day celebration. Photo: CALTRANS

Thousands of people attended the double dedication ceremony held at the bridge site. Also being celebrated was the opening of the first hard surface highway, between Tamalpias and the Marin beaches (Highway 1). After the bridge dedication, a parade of hundreds of cars proceeded over the new highway to Stinson Beach, for a picnic.

Except for the concrete and steel lift span and piers, the bridge was made entirely of redwood, hence the name, "Redwood Bridge." Over two million board feet of lumber was used in the construction of the 2477-foot roadway and long timber trestles. A 56-foot plate girder lift section near the northern end of the bridge provided a 40-foot clear channel. The bay west of the bridge is navigable only at high tide by small craft, and since there were no docking facilities, it was seldom entered. During the 18 year period 1931-1949, the bridge opened only six times, three of which were for dredges of the U. S. Corps of Engineers.

This bridge was replaced in 1956 by the existing high-level fixed bridge.

(N.P.C.R.R.) Strawberry Point

From 1873 to 1884, a 4,000-foot railroad trestle bridge spanned Richardson Bay from Strawberry Point to Pine Point. The bridge belonged to the North Pacific Coast Railroad, a narrow gauge line that ran from Sausalito to Cazadero, via Larkspur, San Anselmo, and Point Reyes Station.

Due to a complaint from S. R. Throckmorton, who claimed the trestle impeded navigation to his Rancho Saucelito, the railroad had to install a moveable span (type unknown) in the trestle. Twice a day a locomotive went out to the drawbridge: first to drop off the bridgetender and later to pick him up. A tender was required to be there to open and close the bridge for the occasional schooner that passed through.

In 1884, the tracks were rerouted via a tunnel near Mill Valley. The trestle was later removed.

CORINTHIAN INLET

Belvedere-Tiburon

The question as to whether Belvedere and Tiburon were one land mass or two separate islands was the subject of dispute and litigation for several years in the late 1800s. The owners of the Reed Grant held that Belvedere was a peninsula of Tiburon and belonged to the Rancho Corte Madera.

When Israel Kashow arrived at Belvedere (then called Still Island) in 1855, he liked what he saw and settled down. Calling the Island "Kashow Island" he tried in every way possible, including allegedly digging a trench across the natural causeway (San Rafael Avenue), to make it into an island and thus detach it from the Reed Grant. The dispute over the ownership of the "island" came to an end in 1885 with the declaration that it was indeed part of Rancho Corte Madera Grant.

Although Kashow lost the battle of Belvedere, he came away with deeds to 139 acres of tidelands that included the 20-plus acre lagoon north of Beach Road and much of Corinthian Island.

In 1888 Mr. Kashow built a wooden bascule bridge across the tidal channel that separated Beach Road from Corinthian Island. Until the construction of the bridge, crossings were made either by rowing or swimming.

Raising the drawbridge was always a big event and called for some real manpower at the end of a block and tackle.

Each fall, numerous arks, boats and yachts of all sizes moved from anchorages in Belvedere Cove to the safety of the lagoon. The parade of boats out of the lagoon in the spring marked the opening of yachting season and was the forerunner of today's "Opening Day on the Bay celebration." Hundreds of spectators gathered to watch the lifting of the bridge and the to watch the boats.

This sketch shows the opening of the wooden bascule bridge over Corinthian Inlet c1900 (probably spring judging from the crowd). A community effort, it eventually became an important social event. Photo: Belvedere-Tiburon Landmarks Society.

The chief of police acted as the "official" bridgetender.

In July of 1912, Mr. Kashow's bridge was replaced with a similar structure. A wooden bascule bridge, with a 30-foot draw, it was built as part of the county road. Wide enough for two horse-drawn wagons, it too was manually operated by means of blocks and tackle. One old timer who remembers the bridge said it was known locally as the "Hell Gate Bridge."

While there is no record of how many times a year either of the bridges were opened, it was probably several. It is unlikely that all the boats would, or could, be moved on only a half dozen prescribed days a year.

In the mid to late 1920s work began on the partial filling in of the lagoon. A bulkhead was built near the bridge site and by 1930 a road was built across the former channel. The date of the bridge's removal is not recorded but it was probably soon after the road was finished.

Now landlocked, the lagoon can no longer be entered by boats from Belvedere Cove.

CORTE MADERA CREEK

Corte Madera Creek originates in the Marin County foothills, winds its way down through Ross, Kentfield and Corte Madera, where it is joined by tidal water before entering the bay near San Quentin. During the gold rush years, Marin County supplied San Francisco with fresh water, lumber and firewood, among other things. Whaling ships and schooners tied up at Kent and Ross Landings to load and unload cargoes. The steamers *Tamalpais* and *Ida* were regular visitors to the Creek.

Unfortunately, the lumber business was so good the surrounding hills were soon denuded of trees, allowing the winter rains to send mud pouring down the slopes into the creek. By 1869, the creek had silted to the point where it was difficult for ships to get to Ross Landing. On several occasions, boats were left high, but not so dry, in two or three feet of mud and unable to move. Some were eventually made into permanent homes. Other houses were later built on stilts over the mud, or in shallow water. Together they were the center of controversy off and on for many years. Even today, there are some houseboats and semipermanent homes on the marshes at the mouth of the creek.

This c1928 view looks north through the trusses of the original Corte Madera Creek Bridge. The large gears on the left suggest some sort of mechanical power had been added to the formerly hand operated span. Photo: U.S. Coast Guard.

(Highway 101) Greenbrae* Mile 0.6

The first highway drawbridge to span Corte Madera Creek was a hand-operated, wooden bascule span with timber trestle approaches. It connected to a hard dirt road across the marsh. Built in 1917, the narrow bridge spanned the creek just west of the existing railroad drawbridge and east of the present freeway overpass. Prior to its construction, traffic crossed the creek three or four miles to the west, via a fixed bridge.

After 13 years of hard use, the worn out wooden structure was replaced on August 15, 1930, by a steel bascule bridge. The old bridge was removed, but for several years, the county maintained part of the old timber trestle as a public fishing pier.

Greenbrae Junction as it appeared in 1930. The control house of the "new" Highway 101 bascule bridge over Corte Madera Creek can be seen at top right. The N.W.P. bascule bridge can be seen at the extreme left. Photo: CALTRANS.

The new $157,000 bridge and its trestle approaches measured nearly 1,000 feet, including the 84-foot bascule span. With a 5-foot sidewalk and 44-foot wide roadway, it put a temporary end to the traffic bottleneck caused by the old single lane structure.

Once navigable for at least four miles above its mouth, by 1930, the creek was usually too shallow for any boat large enough to require the opening of the bridge. It isn't surprising, therefore, that there was a period of 7 years when the bridge wasn't opened at all and it was opened only seven or eight times during its entire 30 years of "service." It was removed on July 29, 1960, shortly after the present high-level structure was opened to traffic.

Dudley Key lived in Corte Madera from 1919 to 1929 and remembers riding his bicycle over the single-lane, wooden bridge on his way to San Quentin to buy shrimp. The shrimp were sold off the old ferryboat *James M. Donahue*, beached by the Richmond Ferry slip.

Harry Richards, a lifelong resident of the area and childhood friend of Dudley Key, says he and his friends used to fish from the bridge, which was also used as a wharf. Scow schooners tied up to the bridge and off-loaded bales of hay to farmers who hauled it away to their dairy farms.

The Northwestern Pacific Railroad bridge as it looks in 1995. Now owned by the Golden Gate Bridge District, its future use is uncertain. Maintained in the open position, it has not been used by trains since the 1960s.

(N.W.P) Greenbrae * MILE .6

In 1884, the Tiburon and San Rafael Railroad built a long, timber trestle bridge across the marsh at Greenbrae and installed a small iron bascule span. It originally carried only a single set of tracks, but a second set was added in later years. In the early days, prisoners bound for San Quentin Prison disembarked from the train at Greenbrae Station and were taken the rest of the way by horse drawn wagons.

In the 1920s the creek was still relatively active and old-timers tell of taking turns at the crank to raise and lower the old bridge. "On weekends, the bridge would sometimes be open nearly all day to allow sailboats to pass through." Also in those early years, prisoners bound for San Quentin Prison disembarked from the train at Greenbrae Station and were taken the rest of the way by horse drawn wagons.

The existing bascule bridge, also manually operated, was built by the Northwestern Pacific Railroad in 1924. Now owned by the Golden Gate Bridge District, it is no longer in use. What the Bridge District plans to do with the structure and the N.W.P. right-of-way is not known. Today it stands with its 61-foot leaf raised skyward, conspicuous by the absence of one set of tracks.

The trestle section of the bridge played a bit part in the Clint Eastwood movie, *Dirty Harry.*

This post card, postmarked July 1910, shows two hay scows alongside a dock on what was referred to in the caption as "San Rafael Ship Canal." Photo: Oakland Public Library.

San Rafael Creek is more a tidal canal than a creek, but for a time in the early 1900's, it was optimistically referred to by some as the "San Rafael Ship Canal."

The waterway did have some busy years in the late 1800s, but it shoaled in so badly, boats sometimes had to wait several days for a tide high enough to get in or out on. After being dredged in 1908-1909, two freight steamers, the *Twilight* and the *W.H. Walker,* began daily service to and from San Francisco. It is not known how long the service lasted. Freight service on the "Creek" during 1921, a year for which records are available, amounted to $359,000.

(S.F.& N.P.) San Rafael * Mile 1.8

The San Francisco & North Pacific Railroad built a bascule span across San Rafael Creek in 1878 and again in 1888. It is not known when the last bridge was eliminated, but by the time it should have worn out, two fixed bridges were already in place downstream.

A locomotive of the San Francisco and North Pacific Railroad is shown on the moveable span of this unusual swing bridge over Gallinas Slough, c1900. One end of the span is moved to the right or left by means of the tall boom and bridle. Photo: Fred A. Stindt.

GALLINAS CREEK

(S.F.& N.P.) San Rafael * Mile 2.2

A mostly tidal stream of about three miles length, Gallinas Creek flows into San Pablo Bay via the Santa Venetia Marsh northeast of San Rafael.

In 1879, the San Francisco & North Pacific Railroad built a wooden swing bridge over Gallinas Creek (then Gallinas Slough) about two miles north of the present Marin County Civic Center. While it is unlikely this bridge opened often for water traffic, especially in its final years, the shallow creek did see some activity during the early years. A brick yard located down stream from the bridge was called on by scows from time to time.

In 1911, the original bridge was reportedly replaced by a hand operated, wooden bascule bridge. That bridge was replaed in 1917 by a manually operated, steel bascule bridge (with counterweights) was erected over the creek at the approximate location of the former (wooden) bridge.

Not having been opened in years, the Northwestern Pacific Railroad eventually got permission to remove the drawbridge. By July of 1947, it was gone.

This steel, hand-operated bascule bridge was the last drawbridge to span Gallinas Creek. Located within sight of the present Marin County Civic Center, it had not opened for water traffic in years when removed in 1947. Photo: U.S. Coast Guard.

CHAPTER 5

SACRAMENTO RIVER

The lower Sacramento River was already a busy waterway in 1848, when gold was discovered on the American River near Sacramento. By comparison to the years that followed, however, navigation on the river had actually hardly begun. Within three or four years, hundreds of boats of all sizes were engaged in river commerce. Each year, daring captains ventured farther upriver to serve the new mines. On January 24, 1852, the steamer *Comanche* arrived in Tehama and claimed the honor of completing the longest trip ever made on a California river.

Several places on the river have claimed the title and status of "head of navigation," a distinction usually determined by four factors; the condition of the river at the particular point; size of the vessels; period in time; and the existence and condition of connecting stage roads.

For example, although steamers called at the town of Redding as early as 1854, when Captain Gilmer took the steamer *Belle* to the mouth of Clear Creek, Redding was never a genuine river port, nor head of navigation. Strong currents and treacherous snags, that occasionally tore the bottoms out of boats, kept most vessels of any size from venturing above Red Bluff.

In 1900, a fixed bridge across the river at Red Bluff established that landing as the head of navigation. Their status was threatened only once during the following six decades when, in 1961, the State Division of Highways announced plans for a fixed bridge above Sacramento. While that bridge did become a reality in 1969, pressure from the Tehama County Board of Supervisors and others insured that it was high enough to leave unchanged Red Bluff's claim to "head of navigation" for a few more years.

The period between 1926 and 1942, when the *Delta King* and *Delta Queen* made regular trips between San Francisco and Sacramento, are considered by many to have been the peak years on the river. Competition, first from the railroads, then the highways, caused the eventual demise of river traffic as the major avenue of transportation. A deep-water ship channel eventually opened Sacramento to ocean-going ships, but it never brought back "old fashioned river commerce," nor was it intended to.

(So.Pacific) Tehama * Mile 232.5

One of the first drawbridges to span the river above Sacramento, was built at Tehama in 1871. The cast iron and wood structure, owned by Southern Pacific, measured 800 feet in length. The swing span of this early bridge lasted only a few months, however, as it was swept off its center pier and washed a half mile downstream by the first spring flood. Fortunately for the railroad, it had engines and cars on both sides of the river, so rail service went on relatively uninterrupted. The steamer *Gem*, from Sacramento, acted as a ferry until repairs were completed and the bridge went back into service the following year. With its new swing span in place, the bridge survived until the early 1880s when replaced by yet another bridge. A steel swing bridge, it too had a short life before being replaced by the existing span in 1898.

Although still in use on the company's Roseville-Gerber line, the bridge has been exempt from opening for water traffic for decades and is welded closed. Until the completion of the highway bridge in 1911, this bridge was also host to horses and pedestrians, at their own risk, of course. During construction of the highway bridge, divers found the center pier of the old railroad bridge that had washed away in the flood of 1871–1872.

Above: The 1871 Southern Pacific Railroad bridge at Tehama. The building at left was the bridgetender's residence. Photo: Tehama Library. **Below: The existing S.P. Bridge built in 1898 is still in use by the railroad, but no longer opens.**

(Hwy. A8) Tehama * Mile 232.2

Tehama's first highway, or "wagon bridge," as it was known, was erected at the foot of "C" Street, approximately 35 feet down stream from the railroad span. Built by Pacific Construction Company, it was dedicated on September 19, 1911. The county originally planned to have a swing bridge there, but the inventors of a new vertical lift bridge convinced them to build one of their design. Aside from costing $10,000 less than the proposed swing bridge, it became only the second such bridge in the country. The price agreed upon was $172,963. A steel bridge, with wooden trestle approach spans, it measured 750 feet in length. The 167–foot drawspan, located near the western end of this narrow bridge, provided a maximum vertical clearance of 92 feet.

Not long after this bridge was completed, river traffic slowed to the point where it was no longer a factor. Only an occasional steamer made it that far upriver. The popularity of the railroads eventually ended what water traffic there was, and the last time anyone remembers the Tehama bridges opening for other than maintenance was in 1916.

The Tehama County Road Department reported in 1963 that the highway bridge had not been painted in 43 years. Since they voted against painting it that year also, it likely waited a few more years.

In 1976, a grossly overloaded truck damaged and weakened the structure beyond economical repair. The road department decided to immediately begin the construction of a replacement bridge, the plans for which, were already drawn.

Under severe restrictions, the 65–year old wagon bridge made it until the following year when a fixed bridge was completed a short distance down stream.

The destruction of the old bridge proved an unpopular decision. Many residents felt that, with just the removal of the overhead counterweights, the bridge could have been made safe enough for bicycles, pedestrians and horses. Some felt the bridge might even have qualified as a historical monument or landmark.

One of the earliest vertical lift bridges of its size in the state, the Tehama Bridge, or "wagon bridge," as it was called, carried vehicle traffic over the Sacramento River from 1911 until 1977. The last time it is known to have opened for water traffic, however was in 1916.

(Hwy A9) Squaw Hill * Mile 221.1

Twelve miles below Tehama at Squaw Hill, a steel drawbridge replaced a ferry service dating from 1871. Completed in May of 1922, at a cost of $236,000, this fine swing bridge is named the Woodson Bridge for Mr. W. N. Woodson, proprietor of the Maywood Colony in Corning.

Constructed several years after most river traffic had ended, there is no record of it ever having opened for boats of any kind. With a vertical clearance of 29 feet in the closed position, anything capable of navigating the river at that point could pass under the bridge. In 1961, when the county replanked and asphalt surfaced the bridge's deck, they also filled in the joint between the fixed and movable sections. The alteration essentially made it a fixed bridge. Although still subject to opening by order of the U. S. Corps of Engineers, that wasn't considered likely. Even old-time residents have never seen it open for any occasion; others refuse to believe it ever could.

In the years between the mid-1870s and 1910, Squaw Hill, named for two Indian squaws who lived nearby, was a regular stop for river boats going up and down the river.

The Woodson Bridge shown here was built in 1922 and is still in place. Although capable of opening for water traffic until 1961, there is no record of it ever having done so.

(Hwy 32) Hamilton City * Mile 203.5

The first drawbridge to span the river at Hamilton City was built in 1882. Originally constructed as a "free bridge," the wooden structure cost Butte County, aided by funds from the then Colusa County, $27,000. Not many years after it opened, the center span washed away during a storm. When rebuilt and reopened, it became a "toll bridge." In exchange for the privilege of charging a toll, the Chico Bridge Company took over the operation and maintenance of the structure. The bridge fared no better under the new management than with its predecessor, for in 1908 it caught fire and burned to the water line.

Butte and Glenn Counties joined forces to temporarily replace the bridge with a ferry, which operated from old Chico Landing. Mr. Hiram Crenshaw, who had served as both toll collector and tender on the bridge, was appointed ferryman.

Construction on the last drawbridge at that location, a $300,000 steel swing bridge, was started in 1908, but delayed until 1910, because its center pier washed away during construction. The cost of the new span was shared equally by the two counties. Mr. Crenshaw, who became the bridge's first and only tender, lived on the bridge site in a comfortable cottage provided by Butte County. When water traffic under the bridge became all but non-existent, the county decided the bridge no longer needed a tender.

Christened the Gianelli Bridge, for the Gianelli family who owned and farmed several thousand acres of land east of the river, it served as a vehicle crossing until replaced in 1987. According to records of the California Department of Transportation, it had not opened, except for maintenance, since the state took it over in 1933.

The existing $5.2 million concrete, box girder fixed span at Hamilton City was dedicated in October of 1987. Not long afterwards, the old swing bridge was destroyed in dramatic fashion when explosive charges brought it crashing down into to the river. The City of Chico, located five miles east of the river, traces its beginning to Chico Landing. A regular stop for most steamers until the early 1900s, the landing, originally called Bidwell Landing for General John Bidwell, was at the approximate site of the existing bridge.

The third and last drawbridge to span the Sacramento River at Hamilton City was the Gianelli Bridge completed in 1910. This bridge was no stranger to river traffic and the sound of a steamer's whistle, but in 1987, when it was replaced by the existing concrete bridge, it hadn't opened in at least 50 years. Photo: CALTRANS.

(No. Elect. R.R.) Hamilton City * Mile 203

In 1906, the Northern Electric Railroad began construction on a trestle railroad bridge across the Sacramento River, south of Hamilton City. This was no ordinary drawbridge, however. The drawspan consisted of a pontoon section on which tracks had been affixed to accommodate the rail cars. When necessary to allow the passage of water vessels, the span floated free of the fixed sections and out of the channel.

The Hamilton City branch of the Northern Electric Railroad began service on September 12, 1907, and operated both streetcars and freight cars over the bridge.

The streetcars began the trip from the Interurban Depot in Chico and ran on the local streetcar tracks as far as the Southern Pacific Depot. After crossing the S.P. tracks, they rode Northern Electric rails to the river bank. After crossing the river, via the pontoon bridge, they were back on private rails into Hamilton City. The total distance—11.4 miles.

By 1913, after some interruptions in regular service, and some longer periods of no service at all, the service was reduced to summer months only. In 1915 the section of the route between the S.P. Depots in Chico and Hamilton City was abandoned altogether and the tracks removed.

Flooding and swift river currents were an ever–present problem that often prevented or delayed the streetcars and trains. Whenever the bridge was out, passengers were carried across the river via a cable–ferry. Streetcars from Chico ran to the banks of the river where passengers boarded the ferry. Once on the other side, they transferred to another streetcar and continued on to Hamilton City. Usually, four round trips a day were made.

The beet trains were not so fortunate. When the bridge was unusable, the train couldn't get to the beet factory and back to the Southern Pacific tracks. Fortunately, sugar beets have a rather flexible harvesting schedule, so having a railway connection only part of the time did not necessarily spell disaster.

Although the date is not recorded, the bridge was probably removed in 1915, about the same time, or shortly after the removal of the tracks.

(Hwy 162) Butte City * Mile 169.7

The method of crossing the river at Butte City has alternated over the past one hundred twenty years, between ferries and drawbridges.

A ferry service established in 1875 by the Marysville–Shasta Stage Line was replaced in 1887 by a wooden drawbridge, but not for long. Ten days after it opened, flood waters swept it away, and it was back to the ferry again. A short time later, a second drawbridge rose on the very foundations of the original span. Completed in 1894, it fared only slightly better than the first. Two years after its completion, a steamboat went out of control in flood waters, knocking the bridge into the river.

Incredible as it may seem, the 292-foot moveable span of this bridge, built at Butte City in 1949, was opened and closed by a hand operated capstan, as seen in the photo. By the time the bridge was actually completed, a drawbridge was no longer needed there. It has never opened for water traffic. Photo: CALTRANS

Once again the ferry came to the rescue and for the next ten years served as the only link between Butte City and the area west of the river.

In 1904, an impressive structure with an iron swing span was completed at this location. It looked like the end had finally come for the ferry. The new bridge did serve more or less uninterrupted for some 16 years, but was condemned and closed in 1940. It remained in place until 1946, however, before being dismantled and removed. Anywhere but in Butte City, that might have been a catastrophe, but to the people of that area, it merely mean that it was the ferry's turn once again.

During the periods when the ferry was the only way across the river, farmers of the 39,000 acres of farmland on the Glenn County side and the approximately 10,000 acres on the Butte side suffered great losses. For the farmers, the ferry was a poor substitute for the bridge when it came to delivering their products to the market.

When the bridge was condemned, the state already had plans for a new bridge to be built at that location. In fact, bids had already been sent to contractors with a completion date set for June of 1942.

A ferry and support equipment, costing approximately $25,000 was purchased and operations were to start in May of 1941. The ferry was to run from Princeton until arrangements could be made at Butte City. Unfortunately, due to a shortage of steel during the war, the bridge was put on hold indefinitely.

Finally, the 3/4–mile long bridge, costing over $500,000, was completed and opened to highway traffic on January 17, 1949. Its manually operated 292–foot swing span has never opened for water traffic and, since 1986, has been exempt from opening at all. It was last opened on December 15, 1980 to check its mechanism. With 15–1/2–feet of vertical clearance at low water, there is plenty of clearance for any boat able to navigate in those waters. Boats with tall masts must either lower them or stay down river. With the existing bridge still in good condition after 45 years, it's a good bet the ferry is retired for good. When this bridge is replaced, it will be with a fixed structure.

The state ferry that served the area from May 1941 to January 1949, was moved to Stockton to be used as a spare.

(Hwy 45) Colusa * Mile 143.5

Colusa started as a river town in 1850, with the arrival of John Semple from Benicia. He brought with him a boat load of building materials for the construction of Colusa's first buildings. By 1852, much of the river traffic ended at Colusa. Passengers and freight destined for points north transferred to stages there to continue their journeys. Barges carrying up to 600 tons were arriving in Colusa by 1862.

Colusa was also a transportation hub. In 1874, nine stage lines established tri-weekly service to many parts of California. For a time, competition became so keen, the stage runs were like horse races between the various companies, causing prices to tumble to almost nothing. With the coming of the railroad to the east side of the river in 1872, and to Colusa in 1876, the stage business all but ended. In later years, auto stages became so popular they were able to reclaim some of the old stage routes.

River traffic, which had fallen off considerably, managed to survive, due in part to the fact that the railroad owned many of the river boats. In 1914, four passenger steamers were making the trip between Colusa and San Francisco each day, but by 1917 the number of trips was down to only two. Service continued on an irregular basis until 1925. Colusa, Sacramento, Petaluma and Stockton were the last river ports for passenger service in California. It has been many decades since any sizable boat could navigate the river beyond Colusa.

The first drawbridge to span the river at Colusa was a wooden swing bridge. Built at a cost of $16,000, it was dedicated on December 28, 1881. Although records do not show how long it was in place or what happened to it, it is known that a new bridge occupied the same location in 1900. This second bridge, an iron swing bridge with a wooden roadway, cost the county $50,000. It is not known just what happened to that bridge either, but fire and floods were the usual undoing of wooden bridges over the Sacramento River.

When the last drawbridge at Colusa was removed in 1980, a 99-year history of drawbridges in this city came to an end. The swing span is now on display in the Sacramento River State Park in Colusa. Photo: CALTRANS.

The third, and last true drawbridge to span the river at Colusa opened to traffic in 1908. Twenty-feet wide by 608-feet long, the little swing bridge saw plenty of water traffic during its first decade of service and had its share of mishaps. Also a busy highway bridge, it was burdened for years by height, width and finally weight restrictions. Even though it was given major overhauls in both 1948 and 1971, it was declared unfit for legal loads in 1972; resulting in a motion by the Board of Supervisors to have plans drawn up for a new bridge. Application was made to the Coast Guard for a fixed structure, but the city had to settle for a removable span bridge.

The site chosen for the new bridge was immediately downstream from the ailing swing bridge. Construction began in 1979 and ended in the summer of 1980 with the completion of the approach spans and the lowering of the steel removable span into place.

Soon after the new bridge opened, the fixed sections of the old bridge were demolished. The county salvaged the swing span, which is now mounted on a concrete foundation at the Sacramento River State Park at Colusa. A bronze plaque describing the old bridge is mounted in a stone monument on the river levee at the former location of the bridge. An unusual, but fitting tribute for 72 years of faithful service.

(Hwy 20) Meridian * Mile 135.5

Meridian reportedly was named for the Mount Diablo Meridian which passes nearby. For lack of a more credible explanation, most of the town's residents are willing to accept it as gospel.

An important river landing in the early days, Meridian survived as a shipping point for potatoes and other agricultural products for several years after most other upriver landings had been abandoned.

In 1912, the Northern Electric Railroad contracted with Thompson Bridge Company to build a steel swing bridge at Meridian. Costing the railroad some $24,000, its elaborate finials atop the swing span set it apart from all bridges on the river, except the "M" Street bridge in Sacramento, also owned by the

The old Northern Electric Railway Bridge at Meridian. By the time the above photo was taken, the span was being used exclusively by motor vehicles. The N.E. tracks were removed in c1940 and the state took over the bridge in 1960. Photo: CALTRANS After 64 years of service to trains, autos, horses and pedestrians, the old bridge was replaced in 1977 by the modern cable-stayed swing bridge shown below. Photo: Gordon Cinnamond.

same railroad. The dedication of the bridge on April 1, 1913, signaled the coming of the railroad to Meridian and gave rise to a two-day celebration attended by more than 3,000 people.

On February 3, 1915, less than two years after its installation, flood waters washed away one pier and the west approach span, closing the bridge to railroad traffic for eight months. During that period, "commuters" from Colusa hitched their horses on the west side of the river, climbed a ladder rigged from shore to the bridge, and caught the train at the depot in Meridian.

The railroad bridge, which originally carried only a single set of tracks, was also used by horses, motor vehicles and pedestrians. After the bridge was reopened in 1916, a single wooden roadway was added to each side of the bridge outside of the trusses. In that configuration, it officially served as a combination railroad and highway bridge. In the late 1930s train service over the bridge was discontinued. The removal of the railroad tracks soon afterwards completed the transition from railroad to highway bridge. The state took over the span in 1960.

In 1975–76, the old Meridian Bridge was dismantled to allow for the construction of the existing bridge. Traffic was carried across the river on two temporary bridges during construction.

Completed in 1977, the new cable–stayed, girder swing bridge cost the state $5.3 million. The only bridge of its kind in California, it was designed to lessen the impact on the environment and to try to hold down costs. The old center pier and portions of the abutments were salvaged and used in the new structure. With only 15 feet of vertical clearance, this bridge, unlike its predecessor, does have to open occasionally for barges and pleasure boats with tall masts. Twelve hours notice is required for opening.

During the dismantling of the old bridge, the finials, bell, plaque and horn were saved and donated to the Sutter and Colusa museums. The Sutter County Museum has two finials, a plaque and ducks carved out of wood taken from the deck of the bridge.

The last Southern Pacific drawbridge at Knights Landing as it appeared in 1971. Built in 1904, it carried both trains and autos for many years. This bridge is no longer in place.

(So. Pacific.) Knights Landing * Mile 90.0

The earliest bridge to span the river at Knight's Landing was built by the Southern Pacific Railroad prior to 1872. Not unlike the company's bridge at Tehama, it too washed away with flood waters in 1872. During the reconstruction of that bridge, the ferryboat *Dana* was used to carry passengers from one side of the river to the other.

In 1904, that bridge was replaced by a narrow, thru truss swing bridge. Until the existing highway bridge was completed in 1933, the railroad bridge also carried pedestrians, horse drawn and motor vehicles. In 1960, when it was still tended 8 hours a day during the week and 24 hours on weekends, the bridge opened 386 times. A gas engine provided power to open and close the span. During its last years before being removed, it was maintained in the open position.

(Hwy 113) Knights Landing * Mile 90.1

The double-leaf Knights Landing Bridge opens for the tug *Colusa* and its tow, the oil barge *Placer*. They are heading upriver. Photo: CALTRANS.

A short distance upstream from the former S.P. Bridge, the state built a double–leaf bascule bridge in 1933. Built at a cost of $137,000, its two 111–foot leaves carry a 24–foot roadway and two sidewalks.

The town of Knight's Landing and the entire area went all out to welcome the new bridge at its dedication on December 2, 1933. An opening–day parade kicked off the event, followed by a baseball game and, later in the day, an aerial show featuring a daredevil pilot flying between the raised leaves of the bridge. After years of driving on the railroad bridge, everyone who used the bridge was unanimous in their praise for the new span. In the heart of a vast agricultural area, it is an important highway link.

While not as active as in the past, this bridge still opens several times a month for water traffic. Twelve hours notice is required for opening.

SACRAMENTO

In its role as the "Nile of the West," the Sacramento River played host to more than five million tons of shipping a year during the first two decades of this century. Even as early as 1878, the figure was nearly 200,000 tons. The City of Sacramento, located approximately 60 miles above the mouth of the river, is the largest inland port in California and has been number one in importance since the cry of "Gold!" rang out in 1849. Repeatedly ravaged by devastating floods and fires in those early years, it no doubt owes its existence to men who never knew the meaning of the word "quit." Levees and other flood control measures, and finally Shasta Dam, have since tamed the river to where floods, at least, are no longer a major threat.

Broad Street * Mile 59.2

Early in 1858, the first bridge ever to cross the Sacramento River was authorized to be built as a private toll bridge for a term of 30 years. The Sacramento-Yolo Bridge, which opened to the public on June 27, 1858, connected Broad Street in Sacramento with Ann Street in Washington. Costing $60,000, the bridge measured 800-feet long by 28-feet wide and was, according to reports, "so well built it weathered several floods." The entire superstructure was of wood, having wooden continuous truss approach spans and a wooden through truss swing span. The swing span, supported on 17 rollers, could be opened and closed by one man using a ten-foot iron hand spike. Until a few years before the bridge's construction, an Indian canoe operating at the same location was the only ferry across the river at Sacramento.

Although there is no doubt it was the largest swing bridge built in California by that date, it was not the first, as some have suggested. By 1869 it was gone.

Seen in the background is the Sacramento-Yolo Bridge, the first swing bridge to span the Sacramento River. Built in 1858, it had washed away or burned down by 1869. Photo: Southern Pacific Railroad.

On January 29, 1870, the California Pacific Railroad dedicated a swing bridge on or near the site of the former bridge, possibly using the same piers. The steamer *Belle* served as a ferry during construction.

That structure was completely overhauled by the railroad and reopened as a new bridge on December 6, 1878. It was an unusual bridge in that its roadway was divided into three lanes or sections, the center section for trains, one outside lane for pedestrians and the other outside lane for horse vehicles. The bridge was so busy during that period, watchmen were permanently stationed at each end of the bridge to warn people when the bridge was going to open. Although manually operated at first, it was later equipped with a steam engine to facilitate faster and easier operation.

The railroad came to Sacramento in 1863 and had a profound influence on the city and traffic on the Sacramento River above Sacramento.

The swing bridge in the above photo was built in 1870 by the California Pacific Railroad. It was constructed on the site of the former Sacramento-Yolo Bridge. Photo: Southern Pacific.

"H" Street * Mile 59.6

In December of 1895, Southern Pacific Company and Sacramento and Yolo Counties agreed to share the $261,000 cost of a new bridge at "H" Street. For some reason, what they planned as an iron bridge, ended up as a wooden bridge. The two county governments balked at paying their shares and the case went to court. In the end, Sacramento County paid $30,000, Yolo County $10,000 and Southern Pacific, the balance of $221,000.

The bridge, which carried trains on one level and motor vehicles on another, was the first double-decked bridge in California. It was torn down in 1912, after the present steel bridge was completed at "I" Street.

California's first double-decked drawbridge. Built in 1895, it replaced the above bridge, which by then was owned by Southern Pacific. Photo: Bancroft Library.

"I" Street * Mile 59.5

The "I" Street Bridge, completed in 1912, carries two vehicle lanes and two pedestrian walks on the upper level and two sets of Southern Pacific tracks on the lower. Total length of the bridge and approaches is 2,194 feet, including the 390-foot swing span. A so–called "deadman's curve" on the western approach span made it the most accident prone bridge in the state. Until corrected, it averaged about two fatalities per year, not counting ten people killed in 1949 alone. Most of the problems were eliminated when the approaches were lengthened and straightened in 1937 and again in 1959. Built by Southern Pacific at a cost of $786,000, it is still operated by them and tended 16 hours daily in the summer, 8 hours daily in the winter. Four hours notice is required at other times. All drawbridges across the Sacramento River, from American River Junction to Isleton, are on this same schedule.

At 80 years old, the existing "I" Street Bridge is still active. Owned and operated by Southern Pacific, it carries their trains on the lower deck and autos on the top. Photo: CALTRANS.

In 1915, a big year on the river as far up as Sacramento, 2,639 vessels of various kinds passed through the draw of this bridge, even though it was beyond the terminous of many of the largest vessels. By 1960, the figure was down to 980. Today it is lower still. The completion of the deep water ship channel in 1961 caused a general movement of ship facilities from the center of town to the new Port of Sacramento. Today, nearly all the traffic through the draw at "I" Street is made up of barges, pleasure boats and other small craft, many of which do not require the bridge to open.

"M" Street * Mile 59

In 1910, the Sacramento Northern Railroad received a 50–year franchise (1910–1960) to build and operate a bridge across the river at "M" Street. Construction on the bridge to unite Brodrick and Sacramento began that same year. A combination railroad and highway swing bridge, it carried railroad tracks down the center between the trusses and vehicles outside the trusses.

Designed at a time when the automobile age was in its infancy and built mainly with the railroad in mind, the structure was not up to the crush of traffic that came with the years. It was obsolete for most of the last half of its 25 years of service. In 1915 it recorded 5,161 openings and was still opening on an average of nine times a day in 1930. By the time it was replaced, this important bridge, in the state's capitol, had gained the reputation as the state's worst traffic bottleneck.

Built in 1910 by the Sacramento Northern Railroad, this bridge carried both trains and motor vehicles. The barges of grain are typical of the type of traffic carried on the river at that time. Photo: California State Archives.

In April of 1926, application was made to the Corps of Engineers for permission to build a high–level fixed bridge over the river at "U" Street in Sacramento. It was to be called "Memorial Bridge." A permit was granted and later extended, but construction still hadn't started when time ran out on the permit. Highway 99 now crosses the river, via a high-level fixed span, a few blocks south of where that bridge would have been erected.

It is not clear whether or not this bridge was intended to replace the "M" Street bridge, but in 1934 the "M" Street swing bridge was removed to make room for the construction of the existing vertical lift bridge at that location.

During construction of the new bridge, vehicle traffic was detoured to the "I" Street bridge. Cars of the Sacramento Northern Railroad crossed the river over a temporary structure located 70 feet north of the building site. Carrying a single set of tracks, the temporary bridge consisted of old railroad girders supported on wooden piles. An 80–foot vertical–lift span stood over the channel. The temporary structure cost the state $90,000, including the *rental fee* for the girders. It was not entirely removed from the site until May of 1936.

Opened to traffic on Sunday, December 15, 1934, the $1,000,000 Tower Bridge was the recipient of the award for "Most Beautiful Moveable Steel Bridge completed in 1935." A thousand carrier pigeons carried the announcement of the bridge's opening to cities all over California. The birds were released during the dedication ceremony.

Because the Sacramento Northern Railroad still had 25 years to go on their franchise, the state was obligated to make one important design concession to them. The train tracks were to be located in the center of the span with protective curbs on each side to keep vehicle traffic off the tracks.

When first opened for traffic, the 52–foot wide deck carried two four–foot sidewalks (outside of the trusses) and four narrow vehicle lanes divided by a single set of railroad tracks. The tracks were later removed and the deck was divided into four clear vehicle lanes. The sidewalks were not disturbed.

This c1936 photo shows the *Delta Queen* clearing the Tower Bridge draw on its way downstream. The *Delta Queen* and *Delta King* were, without a doubt, the most famous boats ever to operate on the Sacramento River. Photo: CALTRANS

The 202–foot draw span is lifted with the help of 1,200 tons of counterweights supported by two 200–foot towers. Still in place, but with a high–level freeway structure now carrying Interstate 80 traffic across the river a short distance to the south, Tower Bridge carries only a fraction of the traffic carried by its predecessor.

On September 19, 1987, only five weeks after a deadly accident on the Three–Mile Slough Bridge, three people drowned at this bridge, when their pickup truck plunged into the Sacramento River while the span was raised. The bridgetender, under the influence of alcohol, overrode all of the safety switches that would have activated the flashing lights and crossing arms. There was nothing to stop the truck from driving off the end into the water.

This bridge is subject to the same regulations as the "I" Street span.

(U.P.R.R.) Jefferson Blvd. * Mile 44.2

The deep–water channel that extends from Suisun Bay to Washington Lake in Sacramento was made possible by a federal project completed in 1961. In Sacramento, the channel joins the Yolo Barge Canal with its navigation lock that lifts vessels back up to the level of the Sacramento River, a difference of 12 feet at that point.

When Completed in 1961, this bascule bridge spanned the only navigation lock in the state. Originally intended to serve ships and barges, today it is only for the use of pleasure craft. The bridge is no longer required to open. Photo: CALTRANS.

The channel follows the course of the Sacramento River to Cache Slough, then through a 23–mile excavated cut, for a total of 38 nautical miles. Maintained to a depth of 30 feet, the channel's width is approximately 300 feet for its entire length. It shortened the distance to Sacramento by about 20 miles and, for most vessels, at least two hours. The savings were accomplished, in part, by avoiding four drawbridges and most of the turns in the river. The completion of the project made Sacramento a deep water port.

The barge canal, together with the 86–foot by 640–foot navigation lock is one mile long. Until a few years ago, this was the only navigation lock in California.

On September 15, 1961, the state installed a combination rail and highway bascule bridge over the lock. The pony truss span provides 86 feet of horizontal and 17 and 20 feet of vertical clearance. Although the rails remain on the bridge, the Union Pacific Railroad abandoned the line in 1990.

Also in 1990, there was a push by industry to build a fixed bridge over the lock in place of the moveable span. Because the proposed bridge did not provide sufficient vertical clearance over the navigable waterway, permission was refused by the Coast Guard.

Citing the lack of use of the navigation lock, a U.S. senator managed to have the barge canal and lock taken off the list of navigable waterways. The bascule span is still there, however.

Intended initially for commercial vessels, use of the barge canal and lock declined drastically in the late 1960s. The new Port of Sacramento facilities below the lock are no doubt the reason.

In 1981, the lock was closed by the Corps of Engineers and remained in caretaker status until November 1982, when the City of Sacramento took it over. Operated on a limited basis for pleasure boats only, the city was forced to relinquish it in 1987. Even the exorbitant fees it charged boaters to use the lock were not sufficient, the city just couldn't afford to continue.

It is currently operated by the Corps of Engineers for pleasure boats only. The bridge is no longer required to open for water traffic.

(I–80) Yolo Causeway

Two of the most unusual drawbridge situations in California history were the bascule spans in the 3–mile causeway over the Yolo Bypass southwest of Sacramento. The bypass is an extensive area of low land stretching from north of Rio Vista to south of Marysville. The bypass, together with a vast system of levees, is used as a means of flood control. Whenever the levees are in danger of breaking, water is released into the bypass to relieve the pressure against them. The bypass is sometimes flooded for three months in the winter and impassable for three more. When not flooded, the land is farmed. Until 1916, a dirt trail called Tule Jake's Road served as the only route across the bypass during the dry season.

The state built the first causeway, a two–lane, three-mile long concrete and timber structure, between 1914 and 1916. In cooperation with the Bureau of Reclamation, the highway department installed an 80–foot, plate girder bascule span to allow dredges to pass through in case of an emergency. Built at a cost of $25,000, the electrically operated span was situated near the eastern end of the causeway. The entire project cost the state $400,000.

The Yolo Causeway under construction c1916. The gap in the roadway was left in preparation for the bascule span that was later installed. Photo: CALTRANS.

Although opened to traffic on March 18, 1916, it was dedicated May 11 through May 14 with a four–day celebration attended by thousands. The Yolo Causeway was a badly needed and most welcome project. Until its construction, going from San Francisco to Sacramento by auto during the winter was 120 miles farther than in the summer.

In 1933, twenty feet of roadway was added to the south side of the causeway expanding it to four lanes. The original bascule bridge was replaced by a wider bascule span.

The project, one of the state's largest that year, provided several months of employment to hundreds of men during the Great Depression. During one 45–day period alone, 500 men were employed at hand labor on the structure. In addition, hundreds of men were employed in the depressed lumber industry to cut the three million board feet of redwood needed for pilings and the quarter–million feet of fir used for decking.

Taken in 1929, this photo shows the Yolo Causeway with the bascule span in place. In 1933, when the road was widened 20 feet, a wider span replaced this one. Photo: California State Archives.

The uniqueness of the drawbridges was in their location. Moveable bridges had been built over dry land before, but only when a navigable channel was to have been dredged under them at a later date. The Yolo Bypass, when flooded, was more like a lake than a river or a channel and was not classified as navigable.

Although not opened often, both bridges were opened at least once or twice for the passage of dredges. They were also opened on a regular basis for inspection and servicing.

The existing Yolo Causeway, which carries Interstate 80, consists of a pair of concrete structures without moveable spans.

Freeport * Mile 46.0

Two large riverboats racing for the draw at Freeport c1933. This bridge is still active, opening on the average of nearly 60 times a month. Photo: California State Library.

The Freeport Bridge, located about twelve and one-half miles down river from Sacramento, replaced a flat–bottom ferry that operated at the location for many years.

Prefabricated in 1928 at the Moore Dry Dock Company in Oakland, the bridge's superstructure was towed upriver on barges and erected on concrete foundations the following year. The bridge opened to traffic on New Year's Day, 1930. Although built as a joint venture by Sacramento and Yolo Counties, it has never connected directly to any Sacramento County road.

A double–leaf, heel trunnion bascule bridge, its two leaves account for 226 feet of the 656–foot–long structure. It has only two narrow vehicle lanes and no walks. There was apparently little foot traffic at that point along the river in 1930. Actually, that hasn't changed much.

While the completion of the deep–water ship channel reduced ship traffic on the river, this bridge still opened approximately 500 times for water traffic in 1994.

Freeport was the terminal of the Freeport Railroad, a branch line of the Sacramento Valley Railroad. In 1862–63, many steamers unloaded passengers and freight at Freeport to be transhipped by rail to Sacramento and as far as Virginia City. The Central Pacific Railroad reportedly bought the line.

An overhaul of the bridge in 1955 left it little changed from its original appearance. In 1992, the control house was replaced, traffic lights and gates upgraded and the bridge controllers computerized.

Paintersville * Mile 33.4

The double-leaf bascule bridge at Paintersville is nearly a carbon copy of the Freeport Bridge. Built in 1923 by Strauss Bascule Bridge Company, of Chicago, it is a bit narrower and slightly shorter, however. The only noticeable differences are the concrete abutments, which were originally fitted with handsome lamp posts. Unfortunately, they were removed some time ago.

Courtland, situated just north of this bridge, was a landing for all river boats from the time of its founding in 1870, until the early 1900s. In 1879, a disastrous fire wiped out most of Courtland, which including one of the largest Chinese settlements in the state at that time.

Fruit and vegetable farmers on both sides of the river, above and below Steamboat Slough, took their produce to Courtland where many boats called daily.

The *San Joaquin #2,* with barges *Ohio* and *Serrana* in tow, is seen heading north after passing through the Paintersville Bridge. Photo: California State Library.

Walnut Grove * Mile 26.7

The shortest bridge to span the Sacramento River was the old Walnut Grove Bridge completed in June 1916. Another Strauss bascule span, it measured only 340–feet from end to end. The two 113–foot leaves accounted for two thirds of its length.

Located in the center of Walnut Grove, it was the only bridge on the lower river to provide a single pedestrian walk. The walk may have been an afterthought, however, since it was situated outside of the truss, on the downstream side.

In 1952, the old bridge, plagued with weight restrictions, was replaced with the existing double–leaf bascule span. Until the old bridge was removed, the two nearly identical bridges stood side by side at Walnut Grove. In fact, at that particular time, six double–leaf bridges spanned the lower Sacramento River—plus one across the mouth of Steamboat Slough, parallel to the river. They looked so much alike, rumors had it that a red-hot traveling bridge salesman made a killing in the area.

Standing in the open position, the 1916 Walnut Grove Bridge waits to be removed. Taken in March of 1952, the photo shows the new bridge in the foreground. Built 100 feet upriver from the old span, traffic was not interrupted during construction. Photo: CALTRANS.

Located about 100 feet up river from the 36–year–old span it replaced, the existing span is wider, of heavier construction and carries no weight restrictions. In 1993, the 41-year-old operator's shack was lifted off the bridge and replaced with a new one, complete with the latest safety equipment and controls.

This bridge is tended 16 hours a day during the summer and 8 hours daily in the winter. Four hours notice is required for off hours. Openings average about 1,400 per year.

Isleton * Mile 18.7

The concrete bowstring arch spans of the Isleton Bridge are a refreshing change from the otherwise nearly identical appearance of these Strauss bascule bridges. Opened on October 27, 1923, the same year as the Paintersville Bridge, it is still in service. Each of its two 113–foot leaves can be operated separately by means of two 20 h.p. electric motors. A gas driven generator can provide power in case of an emergency. If all else should fail, three men can open the bridge by hand in about one hour's time.

At least once, the Isleton Bridge did fail to open, and on that particular occasion, the operator of a tug towing a barge downriver was forced to cut the barge loose to save his own boat. The barge passed through the closed draw and suffered only minor damage.

One can appreciate the kind of wear and tear these bridges received, when it is realized that in the first several years of operation, they averaged between 3000 and 5000 openings a year. In 1970, the average was nearly 600 passages a year.

Since 1933, this bridge has been posted for a load limit of 13-tons and a speed limit of 5 and 15 m.p.h. for trucks and autos, respectively.

Established in 1874, and once the center of sugar beet production on the lower river, Isleton also shipped large quantities of wine from her wharves. In the waning years of river traffic, Isleton gained notoriety for its many "houses of ill repute." Like many other bridges along the river, this bridge replaced a ferry service.

The existing Isleton Bridge, judged by many to be most attractive bridge on the river. Photo: California State library.

Rio Vista * Mile 12.8

Rio Vista is one of the "old towns" on the river, having been founded sometime in the 1850s. Like Sacramento, it was plagued with floods that threatened its existence. After a particularly destructive flood in 1861, most of the town literally moved to higher ground. At present, Rio Vista is second only to Sacramento in terms of size and importance on the Sacramento River.

The first bridge across the river here joined the town of Rio Vista with the newly constructed River Road on the eastern shore. Built in 1917, its $260,000 cost was shared jointly by Solano and Sacramento Counties. In 1921 the bridge was taken over by the state and added to the state highway system. The structure consisted of four bowstring arch spans, a 1,600–foot short–span timber trestle and a double–leaf bascule span.

Located just above the mouth of the river, it was a busy bridge from its very beginning and endured its share of mishaps. One such incident occurred in May of 1926. While in the process of opening one Sunday evening, the main drive shaft in one of the leaves broke. The mishap stranded the leaf in the open position and closed the bridge to auto traffic for about 36 hours.

The shaft and gears had to be removed and taken to San Francisco where the gears could be fitted onto a new shaft. Once fitted, they were rushed back to the site and replaced. Working around the clock and using their personal vehicles to transport the parts, workers had the new shaft installed and traffic flowing again by 3 a.m. Tuesday morning. Making the job more difficult was the need to do much of the work from boats and at night. All of the men were commended for their selfless dedication.

In 1927, the bridge was closed to vehicle traffic for several months to replace the long timber trestle which was partly washed away by flood waters. A detour bridge was built across Wood Island to serve during the repairs. Wood Island itself was later removed by dredging.

Built in 1917, The old Rio Vista Bridge served for twenty-nine years as the lowest crossing point on the Sacramento River. From 1946, when the existing bridge, minus its moveable span, was "completed," until 1959, when it actually was completed, the old and new bridges stood side by side. Photo: California State Library.

The existing Rio Vista Bridge, shown from downriver. Dedicated on April 1, 1960, it cost the state about $4,000,000. Photo: CALTRANS.

The reconstructed bridge consisted of 41 forty–foot "A" frame timber spans supported on pile bents. To keep the cost of the project down, some of the decking from the old structure was lifted off and used on the new bridge. The work was completed in March of 1928 at a cost of $70,000.

Richard E. Brown, a former resident of Rio Vista tells of an incident he witnessed one summer. "One time during the summer of 1945," he recalled, "The day was very hot and after the bridge opened, it expanded and wouldn't shut. They had to bring a welder out and remove some metal from the end of one of the leaves, maybe both. Fortunately, no fire alarms came in during that time from across the river." The Rio Vista volunteer Fire Department was (and still is) responsible for much of Brannan Island and all of Sherman Island.

The state spent $185,000 on the old bridge between 1928 and 1943, but it still ended up at the top of their list of bridges needing to be replaced. Although America was at war, or perhaps because we were at war, construction on a new bridge started in 1943. As it is today, the crossing was the only one south of Sacramento connecting two major north and south highways.

The eastern portion of the new bridge, consisting of seven 180–foot truss spans built on concrete piers, was completed in 1945. A vertical lift bridge, it would have its moveable span over a new navigation channel, 575 feet east of the original channel. The long spans were designed to eliminate the accumulation of debris that had plagued the old bridge.

Due to a shortage of steel for the lift towers and lift span, the bridge was not actually completed until 1959. However, until that time, crossovers between the two bridges allowed them to be used as one.

Declared finished, the "new" Rio Vista Bridge opened to the public on January 12, 1946. From 1946 until 1959, when the lift section was completed, the old bridge and the unfinished new bridge stood side by side. The crossovers took advantage of the finished sections of the new bridge and the remaining sections of the old bridge that included the moveable span..

Dedicated on April 1, 1960, the total cost of the finished structure was approximately $4,000,000. It garnered the Steel Institute's award for the "Most Beautiful Moveable Steel Bridge" completed in 1959.

The Rio Vista bridge as it looked on the morning of January 16, 1967 after being hit by the Italian steamer *Ilice.* It was three weeks before even emergency traffic could use the span. Damage to the structure came to more than $500,000. Photo: CALTRANS.

The 306-foot lift span, weighing 1,550,000 pounds, is supported by two 210–foot high towers. Two 30 h.p. motors aided by 1.5 million pounds of counterweight, provide the power to lift the span. When in the raised position, the bridge offers a 270–foot navigation channel with 135 feet of vertical clearance. The bridge's 26-foot wide roadway has two lanes for autos and two pedestrian walks.

All traffic on the Sacramento River north of Rio Vista must pass under this bridge, making it the busiest on the river. Besides towed barges and pleasure craft, as many as five ocean–going vessels a day pass through the draw. Twenty–five or more openings during an eight–hour shift are not uncommon during the summer.

The Rio Vista Bridge is a 24–hour communication center that serves the highway patrol, sheriff's office, Caltrans and the local fire and police departments. Because it is the only bridge on the Sacramento River tended 24 hours a day, its bridgetenders are sometimes called upon to coordinate special bridge openings at other bridges.

On January 15, 1967, the Italian freighter *Ilice* marred the beauty of the bridge when it missed the draw and rammed the fixed span adjacent to the east tower. When the ship finally stopped, approximately 100 feet of the roadway was at the bottom of the river. It was three weeks before the bridge was back in limited use. Damage to the structure exceeded $500,000.

During the reconstruction of the bridge, a surplus LST, borrowed from the Navy, carried pedestrian traffic across the river. Automobile passengers were not so fortunate; they had to go the long way around, via the Ryer Island Ferry. The residents of Brannan and Sherman Islands turned to Isleton and Antioch for fire protection until the bridge reopened.

AMERICAN RIVER

Best known for its role in the discovery of gold in California, the American River was also a navigable river of minor importance during the late 1800's. The 10 h.p. *Etna,* the first steamer of record to ascend the "Rio de Los Americanos," made the trip in 1850. During the flood of 1860, another steamer made it up to Patterson's Ranch, approximately 10 miles east of Sacramento; reportedly the highest point ever reached by a steamer.

The Jiboom Street Bridge in Sacramento is the only drawbridge to ever span the American River. Most people are unaware that the center span of this bridge was once moveable. A close look at the center pier, however, will reveal the many rollers that allowed this span to revolve 180 degrees to permit the passage of boats.

Sacramento (Jiboom Street) * Mile .1

The only highway drawbridge to ever span the American River is the existing structure at Jiboom Street in Sacramento. Erected in 1931 it was named "The Garden Highway Bridge over the American River." The center swing span was constructed in Alameda in 1926, where it served until 1928, as part of the last Webster Street Bridge over the Oakland Estuary.

Accustomed to opening an average of fifty times a day in Oakland, it has not opened fifty times during its entire stay at its present location. It opened for the last time in 1936. It has been exempt for decades. Just a few yards away is a high level fixed bridge carrying an interstate freeway across the river.

With two narrow vehicle lanes and two sidewalks, the cost of this bridge, including the used 351–foot swing span, was $240,000.

FEATHER RIVER

Built in 1917, this was the last of only three known drawbridge to span the Feather River. Located at Nicolaus, it was washed out in 1928 and replaced by a fixed bridge. Photo: California State Library.

For awhile in the 1850's, Nicolaus claimed head of navigation on the Feather River and, like Alviso in the South Bay, was destined to become a "place of great importance." Were it not for hydraulic mining, which filled in the deep holes and eventually made the river too shallow for safe navigation, Nicolaus just might have made it; well maybe not great, but probably a place of importance.

Regular steamer service between Nicolaus and Sacramento began as early as 1850, when the Feather River became one of the main water routes to the "diggins." By 1857, fast riverboats, drawing up to seven feet of water, were carrying hundreds of passengers as far north as Marysville and Yuba City, via Nicolaus. Not long after 1857, however, the river became so clogged with silt, transportation on that waterway all but ended. "Regular steamer service" came to mean anything more than one round trip by the same company.

The river was first bridged in 1870, and, although the exact site is not known, it must have been at a very narrow spot on the river because the structure is reported to have been only 398 feet in length. Built by an Oakland firm, the little swing bridge cost its owners $18,000.

Nicolaus * Mile 9.3

The second bridge of record on the Feather River was a steel swing span built at Nicolaus in 1917. In place for only 11 years, it was washed out in 1928. Flood waters of earlier years had so changed the location of the main river channel, however, a drawbridge at that location was no longer needed.

A railroad swing bridge was reportedly still in place at Nicolaus in 1941, but there seems to be no record of when it was built or removed.

TULE SLOUGH (WINCHESTER LAKE)

Prior to 1927, the Sacramento Northern Railroad erected a lift bridge over Tule Slough (Winchester Lake) about three miles south and west of Freeport. It is unclear, however, whether it was a vertical or bascule lift span and the accompanying photo is of little help. Although it looks like a wooden double-leaf bascule bridge, that possibility seems very unlikely. Since the slough was not considered navigable at the time, one has to wonder why the company built a moveable bridge there at all. Just how long it was in place and whether or not it ever opened for water traffic is also unknown.

This unusual looking structure was erected over Tule Slough on Sacramento Northern's line between Sacramento and Oxford. Photo: U.S. Coast Guard.

With all Sacramento River bridges above Meridian exempt from opening for water vessels, Colusa, with its removable span bridge, is now, more than ever, head of navigation on the river.

Only one bridge, the Rio Vista Bridge, is still tended 24 hours a day. The Isleton, Walnut Grove, Paintersville, Freeport, Tower and "I" Street bridges are all tended 16 hours daily in the summer and only 8 hours a day in the winter; four hours notice must be given at all other times. The Knights Landing and Meridian bridges are untended and require 12 hours notice.

CHAPTER 6

SACRAMENTO RIVER DELTA

MOKELUMNE RIVER – NORTH FORK

The Mokelumne River rises near the crest of the Sierra Nevada Range, flows 140 miles via the North Fork and 145 miles via the South Fork, before emptying into the San Joaquin River 20 miles above its mouth. Although steamers have been known to go upriver as far as Woodbridge, the head of navigation has always been several miles below that point. During the early 1890's, the river saw considerable water traffic, including regular daily passenger service to Benson's Ferry. By 1900, however, nearly all passenger service and most other water traffic ended six or seven miles below there.

Galt–New Hope * Mile 26

Built in c1903, the Galt–New Hope bridge spanned the river at New Hope road, about one-half mile north of Thornton on the Sacramento–San Joaquin County line. The river was narrow at that point, so the bridge's center pier was positioned adjacent to the shoreline allowing navigation on only one side. The extremely lightweight structure consisted of a 164–foot swing span and only five feet of approach span. When replaced in 1970 with a concrete fixed structure, the swing span reportedly had not been opened in more than 35 years. In spite of this, there was a vigorous, but unsuccessful, campaign to stop the government from "blocking" navigation on the river. An upsurge in the popularity of houseboats was largely responsible for the furor.

The construction of the swing bridge in 1903 created a unique situation, in that a fixed bridge was already in place down river at Benson's Ferry.

Benson's Ferry * Mile 23.3

Little is known about the fixed bridge that spanned the river at Benson's Ferry except that it was there from about 1900 until sometime after 1904. It was, of course, an illegal structure, since the Mokelumne was considered a navigable river to above Galt–New Hope.

On November 11, 1910, a steel swing bridge was built at Benson's Ferry to replace the ferry service there. With a thru truss swing span, 60–foot pony truss span and plank roadway, the lightly constructed bridge was posted for maximum loads of eight gross tons. With sufficient water under the south side of the bridge only, it offered a single 65–foot channel.

The bridge was located on Franklin Road, three miles north of Thornton at the Sacramento and San Joaquin County lines. Franklin Road is one of two direct routes between Sacramento and Stockton.

Shown here is the existing bridge at Benson's Ferry, which was completed in 1949. A close examination of the photo will show the old swing span still in place alongside the new one. It is not clear how the new span could open with that arrangement, however. Photo: U.S. Coast Guard.

The weight restrictions on this span imposed serious hardships on trucking in this largely agricultural region, and violators with overloaded trucks took a heavy toll on the structure.

The present bridge at Benson's Ferry consists of a 230–foot steel thru truss swing span and 245 feet of concrete slab trestle. Opened to traffic in 1949, it is much more substantial than its predecessor, but it never rated a power source to open and close the span. There is water only under the south side of this bridge also.

The approach spans of the old bridge were removed shortly after the new bridge was opened to traffic. The swing span, however, was left in place creating a sort of optical allusion. No longer subject to opening for water traffic, it is also inoperable as a drawbridge. The area's utility lines are permanently mounted on the bridge trusses. With 20 feet clearance at low water, few boats are inconvenienced by its stationary condition.

Western Pacific's steel plate girder swing bridge near Benson's Ferry. Built in 1950, it reportedly has never opened for water traffic. Photo: U.S.Coast Guard.

(Western Pacific) Benson's Ferry * Mile 24

About one-half mile upriver from Benson's Ferry, the Western Pacific Railroad tracks cross the river on a plate girder swing bridge. Manually operated, it too has only one draw.

The original drawbridge at this site, also a swing bridge, was completed on June 23, 1908. It was replaced by a similar structure in 1915 that survived until 1950. When retired, the vintage bridge had not opened for water traffic in 27 years. The existing bridge, which has probably never opened for water traffic, is now exempt from opening at all.

New Hope Landing * Mile 18

By 1903, New Hope Landing, the recognized head of navigation on the South Fork of the Mokelumne River, was the embarcadero for much of the area's farm products. Two sternwheel steamers of 400 gross tons each called here regularly. During that particular year, upwards of 50,000 tons of freight was carried on the river.

The first drawbridge at that site was a lightly constructed swing bridge completed in June of 1893. Located on Walnut Grove Road, it spanned the river two miles west of New Hope.

In 1905, that bridge was replaced by a similar, narrow, thru truss swing bridge. Its builders reportedly utilized not only the original center pier, but also some of the trusses from the original structure. Although the manually operated bridge carried a narrow (16 1/2 feet) roadway, its use by trucks and trailers up to 15 tons was permitted.

The last drawbridge to span the river at New Hope Landing was built in 1905. Replaced by a removable span bridge in 1960, it reportedly hadn't opened in the past twenty years. Photo: U.S. Coast Guard.

Like many of the bridges built during the first decade of the 1900s, its center pier was made up of six concrete cylinders, with the swing span resting on rim roller bearings. Later bridges went to a single pier and center bearings to support the moveable span.

This bridge no doubt opened more often than the ones above it, but it saw little water traffic from about 1935 on. It was removed in 1960, shortly after the present removable span bridge was completed. The existing bridge is a short distance to the south of the former drawbridge. An order from the Coast Guard is required for opening this bridge.

Miller's Ferry * Mile 12.1

A swing bridge was built across the North Fork of the Mokelumne river in 1904 at Miller's Ferry. Also Located on Walnut Grove Road, it joined Tyler Island with Staten Island. The bridge replaced a fifty–five year succession of ferries at that location. In spite of load and size limitations and misuse, the bridge somehow lasted over fifty years. The miracle is that these relatively flimsy bridges survived as long as they did. The region is entirely agricultural and the bridges carried considerable truck traffic, much of which was grossly overloaded. The load limit on this bridge, for example, was seven tons per vehicle

The bridge, which consisted of a 220–foot swing span and literally nothing else, was removed on April 25, 1955.

Above: The original Miller's Ferry Bridge as it appeared in 1928. Built in 1904, it survived for 51 years. Photo: California State Archives. Below is the existing Miller's Ferry Bridge completed in 1955.

Built by Lord and Bishop Construction Co. at a cost of $322,324, the existing bridge is still only two lanes, but is wider and sturdier than the old one. Rated for all legal load limits, it should last well into the next century. Manually operated until 1963, it now has an electric motor.

Not unlike the old bridge, the 237–foot moveable span makes up nearly the entire structure. Surrounded by lonely stretches of farmland and miles of dusty roads, the bridge straddles the county line between Sacramento and San Joaquin Counties.

A busy bridge during the summer, it averages about 400 openings per month and is tended 9 a.m. to 5 p.m., May 1 through October 31. Twelve hours notice is required for opening at other hours and months of the year. The vertical clearance is 15 feet and the majority of the water traffic is made up of pleasure craft.

This bridge was put out of service for 30 days during the summer of 1986 by an unusual incident. On May 29, a Department of Fish and Game tanker truck, loaded with 100,000 baby Chinook salmon, received permission to dump the fish into the river from the roadway of the bridge. With the truck parked on the swing span, the tender swung the bridge into the open position. According to one witness, "As the workers prepared to release the fish into the river the span began to tilt and sway from the weight of the 20,000–pound load. It looked like a giant teeter–totter. I thought the whole thing was going to end up in the water." It couldn't have happened at a worse time, the summer is the busiest time of the year for both the area's farmers and boaters Many feared the bridge would be closed to both vehicle and boat traffic all summer. Fortunately, an incentive arrangement with the contractor reduced the time to only 30 days. Cost of the repairs was $80,000.

Mokelumne River * Mile 3.0

The largest of the Mokelumne River bridges crosses the river just below the junction of the north and south forks, three miles above its mouth. Opened for traffic on June 1, 1942, it links Andrus and Bouldin Islands, via State Highway 12, approximately eight miles east of Rio Vista. An active swing bridge, it is tended 16 hours a day during the summer months and 8 hours daily during the winter. Four hours notice is required for all other times.

Located as it is, near the mouth of the river and with as little as six feet of vertical clearance on a high tide, this bridge opens often. Although nearly all of the boats on the river these days are houseboats and other pleasure craft, oil and sand barges and barges of agricultural products still are seen on the river.

The bridge's total length is 1,436 feet, 287–feet of which is the swing span. An interesting idiosyncrasy of this bridge is the need for counterbalancing the east side of the swing span. For some unaccountable reason, the swing span was not equally balanced.

The Mokelumne River Bridge on Highway 12, east of Rio Vista. The busiest bridge on the river, it opens often for water traffic, especially in the summer.

GEORGIANA SLOUGH

Tyler Island * Mile 4.5

About 400 yards north of the Mokelumne River Bridge, Georgiana Slough takes off in a northerly direction for approximately 24 miles to where it joins the Sacramento River at Walnut Grove. For many years, Georgiana Slough was a favorite water route between the San Joaquin and Sacramento Rivers. Now, however, Three Mile Slough, which enters the Sacramento River below the entrance to the deep water ship channel, is preferred by many boaters.

Judging from the number of yearly openings (2,900), however, the route is still popular.

The Tyler Island Bridge is locally known as "Eddie's Bridge" for bridgetender Eddie Peterson who has operated the span since 1947. Although not noticeable in this open position photo, one side of the moveable span is 17 feet longer than the other.
Photo: Lisa Mader.

From Isleton, Route 160 (River Road) goes east for about one mile, crosses Georgiana Slough, and connects to a levee road on the other side. Until 1940, traffic on the road was carried across the slough by means of a small ferry. In that year, a steel truss, rim bearing, swing bridge was built to take the place of the ferry.

This bridge is a little unusual in that its swing span is 17 feet longer on one side than the other (88 feet and 105 feet). The short side is away from the navigable channel and contains counterweights to make up for the difference in length. Although electrically operated, it is said to be one of the slowest opening and closing spans on the delta. When open, the bridge affords 90 feet of horizontal clearance on the long side only. Tended 16 hours daily in summer and 8 in the winter, it requires four hours notice at all other times. The bridge is posted at 5 m.p.h. for vehicles over 5 tons.

In 1983, this bridge was the scene of a boating accident that claimed three lives. At 8:20 a.m. on December 29 of that year, a rented houseboat piloted by an out–of–the–area man crashed into the closed bridge. The craft overturned drowning three people.

The boat operator was apparently not aware that the bridgetender's shift did not start until 9:00 a.m. By the time he realized the bridge would not be opening for him, he couldn't prevent the accident.

In the summer of 1991, a case of indigestion, mistakenly thought to be heart attack, caused the driver of a utility truck to crash through the bridge's traffic barriers in an attempt to rush a fellow worker to a hospital. The bridge had just closed but had not yet lifted the barriers. There were no injuries.

The Tyler Island Bridge is known locally as "Eddie's Bridge," for Eddie Peterson who has been the bridge's tender for 48 years. Eddie and his wife, who also tends the bridge part-time, are well known in the delta. Before becoming bridgetender, Eddie worked on the Twitchell Island Ferry. On April 17, 1991, a bronze plaque honoring him for his many years of service was placed on the bridge.

No longer a landmark along Georgiana Slough, this former Southern Pacific Railroad bridge served for at least 50 years before being removed. A Scherzer bascule bridge, it was rare sight in the state.

(So. Pacific) Tyler Island * Mile 5.7

A mile or so north of the Tyler Island Bridge, one set of Southern Pacific tracks used to cross the slough via a steel Scherzer bascule span. Its 125 foot leaf, which was maintained in the open position, towered above the surrounding area and could be seen for miles. Built in 1929, it was tended one shift a day until 1960, a year in which it opened 177 times. The route has since been abandoned and the bridge removed.

Walnut Grove * Mile 12.4

Georgiana Slough and the Sacramento River meet at the northern tip of Andrus Island, in the town of Walnut Grove. Isleton Road, which skirts the eastern side of the river between Isleton and Walnut Grove, crosses the slough a few yards below the junction of the two waterways and then continues on through Walnut Grove toward Sacramento.

For 62 years, a flimsy swing bridge carried traffic across the slough and became part of the scenery there. Built during the horse and buggy days (1900), it survived into the age of the diesel truck and trailer. As might be expected, its narrow width (16 feet), coupled with weight and speed restrictions, eventually made it an obstruction to traffic. Located in the heart of a rich agriculture area, trucks and large farm equipment were forced to use other routes. The bridge provided one 80–foot draw, on the west side only.

With the opening of the present Georgian Slough bridge in 1962, the old bridge was retired and later removed. Located 200 feet downstream, the new bridge is more than half again as wide and probably several times as strong as the old one. A modern steel girder bridge with clean lines, it was the recipient of the American Institute of Steel Construction's award for the "Most Beautiful Moveable Steel Bridge completed during 1962." Electrically operated, the 200–foot swing span is also equipped with an auxiliary 25 k.w. generator for emergencies. As with the older bridge, navigation is through the west draw.

Tended from 6 a.m. to 10 p.m., May 1 through October 31 and 9 a.m. to 5 p.m. November 1 through April 30, it averages 1,400 openings a year. Four hours notice is required for untended hours.

In 1994, the traffic warning flashers and gates were upgraded and pedestrian railings were added.

Built in 1900, this lightweight swing bridge carried traffic across Georgiana Slough, at Walnut Grove, for 62 years. As trucks got wider and heavier, they had to be diverted to other routes, making this bridge obsolete. Photo: California State Archives.

The existing swing bridge, which replaced the above structure, won the award for the "Most Beautiful Steel Moveable Bridge completed in 1962." A busy bridge, with only one draw, it averages about 1,400 openings per year.

SNODGRASS SLOUGH

(So. Pacific) Mile 3.5

In January of 1912, the Sacramento Northern Railroad completed construction on a swing bridge over Snodgrass Slough, approximately 3 1/2 miles north of Walnut Grove. The slough was navigable for small vessels for nearly six miles above its junction with the North Fork of the Mokelumne River. At one time, Snodgrass Slough joined the Sacramento River near Green's Landing.

Although originally equipped with a steam engine, the swing span was manually operated for most of its years. Last opened in 1958, it was removed sometime in the 1970s. This bridge had water under the left side of the span only.

Built in 1912 by the Sacramento Northern Railroad, this steel swing bridge was in place for at least six decades before being removed. Carrying a single set of tracks, it was a link in the company's Sacramento to Isleton branch line. Its use by the railroad ended long before it was removed, however. Photo: U.S.Coast Guard.

Twin Cities Road * Mile 4.4

About a mile above the railroad bridge, on Twin Cities road, a 1038–foot, $44,966 highway bridge was constructed across the slough in 1931. Because it was in place for at least two years before the highway was completed, it was dubbed the "Orphan Bridge," a name that stuck with it for several years. In 1965, the 183–foot swing span was rebuilt and its original timber trestle approaches replaced with concrete approach spans.

Twenty–four hours notice is required for daytime openings of this bridge; it does not have to open at night. Latest figures show it rarely opens for other than its annual test opening. In 1964, this bridge had opened for water traffic only twice during the preceding 17 years. Its 20 feet of vertical clearance at low tide is apparently adequate for boats using that section of the slough.

Sacramento County has appealed to the Coast Guard to change the notification time from 24 to 72 hours. It's not that they're constantly being bothered by requests to open the span, but because even one opening a year is expensive. Four employees are required to lubricate the gears, jack the bridge up off the locks and turn the manually operated crank that opens the swing span. Because the bridge has no warning signs, lights or barriers, an additional four employees are needed to block off the ends of the bridge for up to two hours. It is hoped that the longer notification time will further discourage recreational openings.

A bridge was approved for this location in 1902, but that bridge was apparently never built.

Located north of Walnut Grove on Twin Cities Road, this highway bridge was completed in 1931, but did not opened to traffic until 1933. Rarely forced to open for water traffic, it once went for 17 years with only two openings. Photo: CALTRANS.

LITTLE POTATO SLOUGH

Terminous * Mile 0.1

From 1927 to 1961, a spur track of Western Pacific Railroad ended at the town of Terminous. Agricultural products of the area were brought there to be cleaned, processed and packaged. In a good year, as many as 20,000 box cars were loaded and shipped from Terminous.

In 1936, the state built a small,

The original Little Potato Slough Bridge as it appeared in 1972. Located near Terminous, on heavily traveled State Route 12, it was a busy bridge for most of its 55 years. Below: the new swing bridge at Little Potato Slough, now called "The Potato Slough Bridge," was completed in 1991. With more vertical clearance than the old span, it is not required to open quite as often. Photo: CALTRANS.

steel, swing bridge over Little Potato Slough a few yards below its junction with the South Fork of the Mokelumne River. The lightly built bridge, which crosses the waterway at an angle of 85 degrees has a generous 24–foot roadway and two 3 1/2–foot sidewalks. A state–operated ferry at that same location ceased operation with the completion of the bridge. San Joaquin County still operates a cable ferry, the *Staten Island Ferry,* across the Mokelumne River close by.

Located 4 1/2 miles east of the Sacramento County line, the bridge served as part of State Route 12 for 55 years. Before being replaced, it was nearly worn out by the passage of up to 13,000 vehicles a day and 300 openings a month during the summer.

In May of 1991, the aging structure was replaced by a new swing bridge. The first moveable bridge built in the delta in more than twenty years, the concrete structure with a 310–foot swing span, carries a 42–foot plus roadway and two sidewalks. Built just north of the old bridge, the two were side by side until the old span was destroyed and removed. The new bridge provides a clear 100–foot channel on each side of the center pier. Although no longer an important shipping point, Terminous is still popular with boaters and route 12 is a busy artery.

The new bridge is tended 16 hours a day from May 1, to October 31. At other times four hours notice is required. With as much as 37 feet of vertical clearance at low water, compared to only 14 feet on the old span, it is not expected to open nearly as often as its predecessor.

STEAMBOAT SLOUGH

Grand Island * Mile 11.2

The Grand Island Bridge over Steamboat Slough. Located on State Highway 160, near Courtland, this bridge, built in 1924, replaced a swing bridge erected in 1905. The original swing bridge has since served at two other locations. Photo: CALTRANS.

Covering approximately 90 square miles, Grand Island is one of the largest islands of the Sacramento Delta. Situated between the Sacramento River and Steamboat Slough, it stretches from Isleton on the south to a point just below Courtland. The island is a vast expanse of agricultural land, interrupted only by an occasional farm building.

Highway 160 follows the river along the west bank, from the Paintersville Bridge to the Isleton Bridge, and crosses the mouth of Steamboat Slough over a double-leaf Strauss bascule span built in 1924. This bridge is a carbon copy of the aforementioned bridges, which were built just a year earlier. The slough is a short–cut between Rio Vista and landings north of Courtland, and saves eight miles by avoiding the large bend in the river near Walnut Grove. In the 1930s this bridge averaged nearly 1,200 openings a year. Today, with four hours notice required for opening this bridge, except during the summer when it is tended 16 hours a day, the number of bridge openings is probably lower than in the past.

The original bridge to occupy the site was a swing bridge constructed in 1905 to replace a ferry service begun years earlier. After the existing bridge was completed, the center pier and approaches of the old bridge were destroyed. The swing span, however, was cut in half and put in storage.

In 1937, the 335–foot swing span was taken out of storage, welded back together and barged to Three Mile Slough. There, it became part of the first Three Mile Slough Bridge. With 10–foot less vertical clearance at its new location, it was even busier than at its original site.

SUTTER SLOUGH

Courtland * Mile 6.4

Sutter and Merritt Islands lie west of the Sacramento River and are separated by Sutter Slough, which enters the Sacramento River a half-mile above the Paintersville Bridge.

In 1939 a small thru truss bridge with a 183–foot swing span was erected across the slough. Until the slough was bridged, Merritt Island was inaccessible by road and the agricultural products of the island were ferried to Hood or Courtland to get them to market. Unlike some others, Sutter Slough was never widely used by boats. Records show the bridge has opened less than a dozen time since its completion.

A big improvement over the ferries it replaced, the Sutter Slough Bridge contributed enormously to the development of the area. While still in place, this bridge no longer functions as a drawbridge.

MINER SLOUGH

Ryer Island * Mile 5.5

In 1933, the state built the existing 367–foot thru truss swing bridge across Miner Slough at the north end of the island. By using the Ryer Island Ferry and this bridge, highway traffic between Rio Vista and Sacramento can stay on the west side of the river and save time and miles over the state highway route. It is located 12 miles north of Rio Vista.

The 191-foot swing span is operated by a one cylinder gas engine and provides a 78–foot navigable channel under the south side only and 22 feet of vertical clearance at low water. Twelve hours notice is required for opening this bridge.

Ryer Island

For nearly twenty years, a private pontoon drawbridge owned and maintained by Sakata Brothers Inc. (Prospect Farms) also spanned Miner Slough, between Ryer and Prospect Islands. Constructed in 1947, it consisted of two pontoon barges fastened end to end with a single–lane wooden roadway down the center. Each barge was 111 feet by 33 feet by 8 feet, making the bridge 222 feet long by 33 feet wide.To open for the passage of boats, one end was released and allowed to swing downstream against the shore.

The structure could pivot from either side, depending on which way the tide was running. The bridge was used extensively for agricultural hauling and was operated by a company employee, who was usually close at hand. Whenever possible, it was left in the open position, especially on weekends during the summer when water traffic was the heaviest. In the open position it afforded unlimited vertical clearance and a 130–foot clear channel. When closed, however, there was no clearance of any kind.

Miner Slough branches off from Cache Slough in the emergency turning basin, at the lower end of the Sacramento Ship Channel. Together with Sutter and Steamboat Sloughs, they surround Ryer Island. This bridge was removed in 1966.

This swing bridge over Miner Slough is located at the north end of Ryer Island. Built by the state in 1933, it shortened the travel time and distance between Sacramento and Rio Vista. It also helped farmers get their agricultural products to market. Photo: CALTRANS.

This double-pontoon bridge over Miner Slough was built in 1945 by Sakata Brothers Inc. Joining Prospect and Ryer Islands, it provided a more convenient route for the shipment of the company's agricultural products. This bridge was removed in 1966. Photo: U.S. Coast Guard

THREE–MILE SLOUGH

Rio Vista * Mile 0.1

Three–Mile Slough is strategically located in the heart of the delta, adjacent to a popular state recreational area and the Sacramento River. Hundreds of boating and fishing enthusiasts use the slough as a short cut between the Sacramento and San Joaquin Rivers, marinas, and fishing grounds. Tugs pulling barges of sand and gravel use the slough year-round and barges laden with sugar beets and other agricultural products use the slough during harvest seasons.

The first bridge to span this waterway utilized the swing span from the original Grand Island Bridge over Steamboat Slough. Built in 1905, the swing span was salvaged when the aging structure was replaced in 1924 by the existing Grand Island Bridge. After thirteen years in storage, the span was moved here in 1937. A very rugged structure and a good traveler, the swing span did not end its career here. After the present lift bridge was completed in 1950, the swing span was moved again to yet a new location. Erected on a new center pier, it now spans Middle River, between Bacon Island and Lower Jones Tract.

Standing in the open position alongside its replacement is the swing span of the "old" Three Mile Slough Bridge. It was later moved to Middle River, where it is still in operation. Photo: Richard E. Brown.

Three–Mile Slough's new vertical lift bridge, with a 175–foot lift span, supports a 26–foot roadway and two sidewalks. Dedicated on August 25, 1949, it is among the most active bridges in the state.

Drawbridges and/or their protective fenders are often the victims of passing boats, but in many cases the incidents go unreported. High tides, strong currents and misjudgments by the boat's operator, all contribute to accidents. Unless the tender is on duty or the damage to the boat is crippling, however, many owners leave the scene to avoid an investigation and possible penalty.

This bridge was the scene of a serious accident on August 9, 1987. Four people drowned when their car plunged off the bridge and into the water. The car, carrying five persons, somehow made it half–way onto the lift span, just as it started to raise. According to a newspaper reports, relatives of the victims charged the deaths were caused by the inability of the bridgetender to see cars on the moveable portion of the bridge. The state, while admitting that monitoring equipment on the bridge could have been better, claimed the driver of the car should have moved forward on the span.

On January 26, 1970 a less serious accident occurred when the bridge was struck by a sand barge under tow of the tug *Tyee* out of Vancouver, Washington. The bridge suffered only minor damage and no one was injured, but the barge didn't fare so well.

This bridge is tended 24 hours daily.

The existing Three Mile Slough Bridge is shown in the partially raised position to allow the passage of a small boat. Vertical clearance in the closed position, during high tides, is only 7 feet. Photo: Richard E. Brown.

CHAPTER 7

SAN JOAQUIN RIVER

Before the railroads came to the Central Valley in 1870-71, steamers were a familiar sight on the San Joaquin River as far up as Hill's Ferry, 118 miles above the river's mouth. During the months of April, May and June, smaller flat-bottomed boats carried on extensive trade up to within 10 miles of Fresno. Shallow-draft steamers also plied the waters of the Merced River for several miles and the Tuolumne river to Empire City.

In those early years, wheat was king in the valley and more than a million tons of the golden grain was moved on the river each season. Barges with a capacity of nearly 20,000 sacks of the grain and drawing as little as five feet of water were commonplace on the river.

Riverboats, some carrying up to 50 passengers in cabins and staterooms, operated regularly on the San Joaquin in the 1860's, a few going as far as Fresno City on Fresno Slough. Although an occasional large steamer would venture above Stockton, navigation even to Hill's Ferry was sometimes a gamble. Most boats on the river above stockton were under 100 tons. The largest vessel ever to reach Hill's Ferry was the 400 ton *Centennial,* in 1877. It took on 6,000 sacks of wheat and made it safely back to Stockton. It never tried it again.

In spite of its apparent popularity, the river failed as a good means of transportation and actually hastened the coming of the railroads, which, in turn, helped bring about the end of navigation on the river.

The main reason river transportation was unsuccessful can be summed up in one word: Irrigation. Each time a new irrigation canal siphoned water from the river, less water remained for navigation. When navigation became impossible, as it sometimes did just at harvest time, farmers used the failure as justification for taking additional water for irrigation. Eventually the irrigationists won the battle. River transportation on the upper San Joaquin River came to an end.

The railroad wasn't all everyone had hoped for either. In 1890, for example, one could ship certain commodities from England to San Francisco, via New Orleans, cheaper than from Bakersfield to San Francisco by rail.

In 1911, in an effort to get the government to consider improvements on the river, the 104-foot steamer *J.R. McDonald* went up river to within a few miles of Herndon. This was the first time in 40 years that a steamer had been up that far, and its return wasn't without difficulty. Ranchers were prevailed upon to divert water back into the river for the steamer's return trip to Stockton.

The government did make a study regarding the feasibility of navigation locks on the river, and Fresno at one time even considered building a canal to bring river transportation to that city. Neither of the plans proved to be practical, however, and were never acted upon.

Firebaugh * Mile 195

When the first settlers came to this region, canoes and ferries were used to get people and supplies from one side of the San Joaquin River to the other. As the population increased and land transportation gained in importance, it became necessary to bridge the river. The first up river ferry replaced by a bridge was at Firebaugh in 1885.

In that year, a 670–foot–long, iron swing bridge was completed across the river where Andrew Firebaugh had established his ferry about 30 years earlier. In 1858, Mr. Firebaugh also built a toll road from Los Banos to Gilroy over the Pacheco Pass. Used for many years by the Butterfield stages, the road become the main route by which the produce of the Central Valley reached San Francisco markets. Miller and Lux, with their more than 600,000 acres, accounted for much of the goods going in both directions over the road. They, more than anyone else, were directly responsible for the coming of the railroad to the west side of the river.

The furthest upriver drawbridge on the San Joaquin River and the first to replace a ferry, was the "Great Drawbridge at Firebaugh" Built in 1885, it served until 1948. Photo: CALTRANS.

The bridge was known for miles around as the "Great Drawbridge at Firebaugh" and each opening was cause for much excitement. In its early years it opened several times a month during the summer, even though it was built after the railroad arrived and river transportation had slowed considerably. There is no record of how many times it opened for water traffic, but during its last 40 or so years, it may not have opened at all.

Condemned several times during its life span, its demise was postponed by frequent repairs, which seemed to make it stronger with age. It never got wider, however, and finally became a traffic bottleneck. As a highway link, it represented an important transportation milestone in the valley and was in use for 63 years. A concrete fixed span built some 200 feet upstream, replaced the swing bridge on June 14, 1948.

The *J.R.McDonald* tied alongside a barge on the north side of the Firebaugh Bridge. Typical of the riverboats that plied the San Joaquin River, it carried both passengers and freight. Photo: U.S. Archives.

This lightly built swing bridge spanned the San Joaquin River at Santa Rita for nearly 35 years. Photo: Lucy L. Travis.

Santa Rita * Mile 175

Twenty miles downstream from Firebaugh, just inside Merced County, another swing bridge was built in June of 1905. Located at Santa Rita, on the east-west route between Chowchilla and Los Banos, it was a much used highway bridge for many years. Built after river traffic at that point had all but disappeared, it is doubtful travelers were inconvenienced often by the bridge opening for water traffic. With 17 feet of vertical clearance at low water, it probably didn't open often, if at all.

In later years, long before the bridge was removed, it was exempted from opening. It was removed and replaced by a fixed bridge in 1939 or 1940.

Dickenson Ferry * Mile 143.7

During the early 1870s, roads in the outlying areas of the county were sometimes fenced across or otherwise altered without any official permission. There were cases where a bridge would be built across a stream, only to find there was no longer public access to it.

While the Dickenson Ferry Bridge, built in 1884 at a place called Chester, may not fit that category exactly, it could hardly be classified as a highly popular or needed bridge. It was claimed that more traffic went through the draw than over the bridge. Traffic on the road was so sparse, regular ferry service across the river at that point was abandoned nearly a year before the bridge was completed.

Built by the Alfred Burell Company at a cost of $8,000, the wooden, manually operated swing bridge was accepted by the Merced County Board of Supervisors on October 16, 1884. Located at a narrow spot on the river, there was water under the east draw only.

This photo, taken from the south side of the Dickenson Ferry Bridge shows the center pier against the western shore The year this bridge was removed is not recorded. Photo: California State Archives.

In July of 1885, the only month for which records are available, the Board of Supervisors paid the bridgetender $60 for opening the bridge six times. Just how long the structure was in service is not known. Today, the eastern end of Dickenson Ferry Road stops 11-1/2 miles from the river. The only sign of what might have been the bridge are some pilings along the shoreline. Chester's "business district," which consisted of a post office, hotel, boarding house and bar, is only a memory and some rubble.

Hill's Ferry * Mile 118.2

In 1852, Jesse Hill began a ferry service across the San Joaquin River at the junction of the San Joaquin and Merced Rivers. A crossroad for hundreds of men going to the southern gold fields in the 1850s, the junction later became an important shipping point for the valley's wheat harvest.

The area came to be known as Hill's Ferry, and its infamous reputation was known far and wide. It readily admitted, or bragged, to being the most "open and lawless" town in the west and the favorite haunt of the famous bandit, Joaquin Murietta and his gang.

When the railroad reached the area in 1888, it bypassed Hill's Ferry and gave birth to the town of Newman, four miles to the west. Actually, Newman was just old Hill's Ferry, transferred to a new location. Everyone and everything, including some of the notorious reputation, moved lock, stock and barrel, to the new town.

Ranchers on the eastern side of the river had been trying for years to gain support for a bridge across the river at Hill's Ferry. With river transportation rapidly diminishing and the railroad on the other side of the river, they had no easy way to get their products to market.

In 1901, after much planning, Merced and Stanislaus Counties agreed to build a drawbridge across the San Joaquin River at Hill's Ferry. Less than a year later, on February 11, 1902, an overflow crowd gathered for the dedication of the new bridge. The occasion included a christening of the span with a bottle

February 11, 1902, "opening day" for the Hill's Ferry Bridge. A large crowd gathered on the west side of the bridge to watch the dedication ceremony and to enjoy a free barbecue. When replaced in 1962, few could remember when it last opened. Photo: Newman Library.

of champagne, followed by a beef barbecue donated and hosted by J.J. Stevenson and Henry Miller. According to newspaper articles of the day, "...such a good time was had by all, that everyone present who had ever been against the building of the bridge, publicly apologized and asked to be 'kicked in the pants.'"

The 472–foot steel bridge, with a 222–foot swing span, was built at a cost of $30,000, which was paid jointly by Merced and Stanislaus Counties. The first tender was Rasmus Lorensen of Newman. Fifteen dollars each from the two counties paid his monthly wages.

Hill's Ferry was the head of navigation on the upper river until well into the 20th century. After the railroad came to the west side, however, few boats of any size passed through the draw here. Late to arrive on the scene, this bridge was never really busy in terms of openings. Its contribution was in joining the two sides of the river.

Considering it served for 60 years, the bridge was truly a taxpayer's bargain. It was replaced in January 1962 by a concrete fixed bridge.

Built in 1887 to replace a ferry service, the Crows Landing Bridge spanned the San Joaquin River for 63 years. Shown in the open position, this photo dates from c1925. Photo; California Archives.

Crows Landing * Mile 107.2

At Crows Landing on May 11, 1887, the Stanislaus County Board of Supervisors accepted a new swing bridge constructed by Pacific Bridge Company of San Francisco. Built to replace Russell's Ferry, it cost the county $36,000. Unfortunately, just three years after its completion, it fell victim to a passing steamer and suffered nearly $5,000 damage. Three years later, in June of 1893, a record high water completely covered the bridge causing minor damage. The span was located approximately four miles east of the present town of Crows Landing.

The original town, a small settlement at a crossroad near the river, received its name and first post office on June 21, 1870. It wasn't many years later, however, before the railroad came to the area and when it did, the town, like others before and after it, moved bag and baggage, to start anew at the railroad.

Among the few who stayed behind was the bridgetender, Mr. Byer. In November of 1889, Mr. Byer managed to have the name of the settlement changed to Byersville and even got a post office again. The post office lasted only until August of 1894 before being moved to the "new" Crows Landing.

When removed in January of 1950, this bridge reportedly had not been opened, except for an occasional inspection, since the 1930s . It was replaced by a concrete fixed bridge.

Grayson * Mile 89.

Southwest of Modesto, Stanislaus County's second drawbridge over the San Joaquin River was dedicated on April 10, 1893. Grayson Bridge, built by Pacific Bridge Co. at a cost of $36,300, ended a ferry service that had operated at that location for more than 45 years. Appropriately enough, the first bridge tender was J.W. Green, the former owner of the displaced ferry.

When built in 1893 to replace a ferry operating near this point, the Grayson Bridge was no stranger to passing steamers, and barges under tow. When removed in 1962, however, no one could remember ever seeing it open. Photo: U.S. Coast Guard.

The original cost of the bridge was only $34,824, but before work could get started, the U.S. Corps of Engineers forced the county to relocate the bridge 260 feet downstream from its planned location. On July 26, 1891, not long before construction was to start, the steamer *A.C. Freese* went aground at the approximate location of the original site, causing the change in plans.

It would seem that the bridge was still built in the wrong place, however, for the water in the river for some time had shown a preference for the route through Laird Slough rather than under the new bridge. Laird Slough branched off to the east above Grayson and joined the river again below it. A dam was built across the mouth of the slough to preserve the original channel, but each winter the dam was partly washed away. By 1911, all signs of the dam had disappeared and the river had changed its course so that nearly all of the water was going through the slough.

When the bridge was removed in 1962, the center pier stood high and dry on a sandbar with little, if any, water on either side. No one could remember ever seeing the bridge open for water traffic. It is known that it once went for 26 years without opening and it may have been much longer. The last time a ship called at Grayson was in 1922, but it didn't pass through the bridge.

TUOLUMNE RIVER

The Tuolumne River, which joins the San Joaquin about two miles north of Grayson, was the scene of brisk water traffic for nearly 20 years in the mid-1800s.

Empire City, the head of navigation on the river, was also a U.S.Army supply station before the turn of the century. The 90-foot streamer *Tuolumne City* provided regular passenger and freight service between Stockton and Empire City, via Grayson and Tuolumne City. Tuolumne City went on to become part of the city of Modesto.

In those early years, land in the area of Tuolumne City and Empire City was sold on the basis of their being near the head of navigation and close to the mines. When the mines closed and navigation on the river slowed to a snail's pace, many of the landowners just picked up and moved on.

Unlike the San Joaquin River, where irrigation took precedence over navigation, the Tuolumne River was ruined by mining operators whose dredges poured silt into the river making it too shallow for steamers to negotiate.

Modesto * Mile 4.O

In 1900, just as river transportation on the Tuolumne River was in its closing days, a swing bridge was built over the river at Tuolumne City, near where Shilo Road crosses the river today. There are indications that the bridge opened for the last time in 1909, which would make 56 years without an opening when it was removed in 1965. Because of its approximate distance from Modesto, it was sometimes called the "Nine Mile Bridge,"

In 1939, the county road department reported the disappearance of the bridge "key", a six–foot long, 80-pound iron bar once used for opening the span. It was feared that some practical joker had stolen the key and might try to open the span. Since there were no warning lights or crossing gates, such a thing could have been disastrous. To solve the problem, the movable span was welded closed, but in such a way that it could be undone if necessary; it never was.

Whitmore's Ferry

A wooden, cable-stayed railroad drawbridge, built prior to 1880, also spanned the river for a period of time at Whitmore's Ferry, south of Modesto. There appears to be no record of what became of that bridge, or when.

Durham Ferry * Mile 72.5

When the first interest in river transportation was to get miners to the gold fields, San Joaquin City was head of navigation on the San Joaquin, and was the jumping off place for the southern mines. The small town, located at mile 74 on the river, was founded in 1849 by Messrs. Titus and Manly who later operated a ferry service there during the 1850s. As the excitement of the mining dwindle, and steamers ventured farther up river, San Joaquin City lost its importance and slowly died.

For a period in the 1870's, its importance was renewed as a shipping point for wheat and other agricultural products of the valley. The future again looked bright for San Joaquin City, but alas, the railroad came to the valley and passed it by. The town had a post office for a few years, beginning in 1852 and again from 1874 to 1888. Each time it was closed for lack of business.

In 1902, when a swing bridge was built near San Joaquin City, it put an end to a ferry service that

Above: The Durham Ferry Bridge, like many others, was named for the ferry it replaced. Built in 1902, it was never a busy bridge in terms of openings. It was removed in 1964. Photo: San Joaquin County Public Works. Below: Southern Pacific's Lathrop drawbridge is shown under construction in 1944. Because of the shortage of steel, the completion of the second lift tower of this vertical lift bridge was delayed until after the war. At the time of this rare photo, the bridge was being operated temporarily as a swing bridge. The single lift tower was being used as a king post to swing the 106-foot moveable span to one side by means of the cables. Photo: CALTRANS

a ferry service that had seen several owners in its time. As was the custom, the bridge was named after the ferry it replaced, in this case, the Durham Ferry Bridge. Like all of the upriver bridges, this one had not opened for many years when removed in c1964.

The wagon road that ran along the west side of the river between Hill's Ferry, Crows Landing, Grayson and San Joaquin City was, in the last three decades of the 19th century, one of the most scenic and exciting roads in the west. In addition to the beautiful pastoral scenery, it abounded with wild life. Literally millions of ducks and geese darkened the skies on occasion, and bear, elk, antelope and even wild horses were a common sight. The ducks and geese were killed by the thousands for the markets in San Francisco, and individual hunters were known to shoot as many as 200 birds in a single day. Over the years, most of the remaining animals were either killed or driven away by fear.

Southern Pacific's Lathrop bridge as it appeared in 1972. When completed in 1944, it was a link in the company's Tracy to Brighton line. It opened to water traffic for the last time in 1951.

(So. Pacific) Lathrop * Mile 56.1

In 1871, Lathrop was a junction of the San Joaquin Valley Railroad and the main trunk route from San Francisco to Omaha.

The forerunner of all drawbridges over the San Joaquin River was the railroad swing bridge built at Lathrop and dedicated on November 10, 1869. It represented a last link in Southern Pacific's route between San Francisco and the east coast. The transcontinental service was actually inaugurated on September 7, but for two months passengers were ferried across the river to waiting trains on the other side.

A natural junction, the area was the site of a ferry service established in 1848 by John Doak and Jacob Bonall. A yawl, built at Corte Madera in Marin County, served as their ferry. Today this same junction is known as the Mossdale Wye; the town of Lathrop is a few miles to the north.

In the days before the Pullman and dining cars, Southern Pacific operated a hotel at Lathrop where passengers could buy meals and spend the night if they wished. Unlike the railroad business, the company didn't have a monopoly on the hotels, so there was lively competition between their hotel and the nearby Travelers Hotel. It was not an uncommon event when the "runners" for the two establishments came to blows. Southern Pacific tried hard to keep the Travelers' runners away from the railroad station, but were mostly unsuccessful.

The original railroad bridge here survived for 29 busy years before being replaced in 1898 by a larger, but similar, swing bridge. Both bridges opened often for river traffic. Even during its last years of service, before being replaced, the latter bridge averaged 113 openings a year, mostly for barges under tow.

The existing Southern Pacific vertical lift bridge, built in 1944, was active until 1951, when it opened for the last time. Two gas engines provided the power to raise the 106-foot lift span.

During the construction of this bridge, the completion of the east lift tower was delayed by a shortage of steel. Until the second lift tower was erected and the counterweights installed, the structure essentially operated as a swing bridge. For a few months, during repairs to the highway span, this bridge also served as a detour for vehicle traffic.

(Western Pacific) Lathrop (W.P.) * Mile 56.7

Built in 1921, this bridge is still in place after 74 years. Intended for the Smith Canal in Stockton, it was placed here instead. It has not opened for many years and is now welded in the closed position.

In 1906, the Alameda and San Joaquin Railroad (now Western Pacific) built a swing bridge a few hundred yards upstream from the Southern Pacific bridge. That bridge served for 15 years before being replaced in 1921 by a newer and heavier span. Although the new bridge was built with the intention of being placed over the Smith Canal in Stockton, the company decided this location was more demanding. While still in use by the railroad after 74 years, it has not opened in decades and is exempt from opening.

The 1906 bridge here was moved to the Smith Canal where it replaced a structure built in 1905, it has since been removed.

Lathrop/Mossdale Wye (Manthey road) * Mile 56.2

By 1900, it became obvious that this area was the logical route for a highway between Stockton and Tracy, so in that year San Joaquin County built a small swing bridge across the San Joaquin River, between the Southern Pacific and Western Pacific bridges. Two railroad swing bridges and a highway swing bridge were now within sight of each other.

A few years after going into service, the relatively new bridge was already being called obsolete. The road attracted so much traffic the bridge was self-destructing.

On August 29, 1927, the state dedicated a new bridge a few yards upriver from the county bridge, which was later removed. A 537-foot steel bridge with a patented 200-foot Strauss bascule span, it cost the state $217,850. Twelve miles west of Stockton and eight miles east of Tracy, the bridge is now part of the Interstate 5 interchange at Mossdale.

The first highway bridge to span the San Joaquin River at what is now known as the Mossdale Wye survived for only 27 years. Lightly built, its popularity with the motoring public was its undoing. It was replaced in 1927 by the existing bascule bridge. Photo: CALTRANS.

Built in 1927, this Strauss bascule bridge replaced the worn out swing bridge. Due to the increased traffic brought about by World War II., it too suffered from over use. During repairs to this structure in late 1944, traffic was detoured across the Southern Pacific bridge. Still in place, this span carries west bound Interstate 5 traffic.

The number and size of vehicles using the bridge increased greatly during the first two years of World War ll. By 1943, the strain on this relatively new structure was showing. Severe load restrictions established earlier were being grossly ignored and congestion was the norm.

Shortage of steel, due to the war, eliminated the option of building a new bridge, so the highway department decided to repair it as soon as possible. A parallel span to carry part of the traffic was to be built as soon as material was available again. In 1944, extensive repairs, including a rebuilding of the 200–foot moveable span, were made on the ailing bridge.

Lathrop/Mossdale Wye (Hwy. 120) * Mile 56.2

It was nearly five years before the long awaited parallel span was opened to the public in June of 1949. The "new," and still standing, 583–foot bascule bridge has a 27–foot roadway and two sidewalks. Its 129–foot leaf, when operational, was lifted by two 15–hp electric motors. A low profile bridge with a pony truss moveable span, it carries east-bound traffic only.

None of the four drawbridges at this important junction have opened for a great many years. Indeed, besides being exempt from opening, all of their machinery has long since been removed.

There is no record of the number of times either of the highway bridges were opened for water traffic, but one can be sure it was not many for the newer one.

Because of the close proximity of the four bridges, the normal signal for requesting an opening was used only for individual bridges. For boats wanting to pass through all four draws, a separate signal was required.

Host to Interstate 5 and State Route 120, the Mossdale Wye is still a major interchange. It is also the site of a popular marina and several permanent houseboats.

When completed in 1949, this bascule bridge brought to four the number of drawbridges within sight of each other at this location. This long-awaited parallel span relieved the traffic pressure on the then existing lone highway bridge. Photo: CALTRANS.

Brandt's Ferry * Mile 45.2

French Camp was the terminus of the Oregon–California Trail used by the French–Canadian trappers during the period 1832 to 1845. In 1844, Charles M. Weber and William Gulnac promoted the first white settler's colony at what is now the site of French Camp and Stockton.

In 1900, a drawbridge was built over the San Joaquin River at Bowman road, connecting French Camp with Roberts Island. Named for the ferry it replaced, the lightly built, thru truss swing bridge survived for 68 years. Heavily laden farm trucks gave the bridge a continuous workout that resulted in many repairs over the years. During the last half of the bridge's life all of the traffic was across the bridge, not under it. It would be difficult, if not impossible to find anyone who remembers the bridge opening for water traffic. It was replaced by a concrete fixed span in 1968.

Garwood Ferry (State Hwy. 4) * Mile 41.6

The original drawbridge at this location, just south of Stockton on Highway 4, was built in 1905 to connect the mainland with Roberts Island. Named for the ferry it replaced, it was literally worn out both by opening and closing and by vehicle traffic. Always in need of repair or undergoing repair, it served until 1933, when it was removed.

The existing swing bridge, completed in 1933, has an electrically powered 258–foot swing span, which makes up all but 43 feet of the bridge. Busy during its first two decades, it has not opened for many years. Both bridges, but especially the original bridge, contributed immeasurably to the development of agriculture on Roberts Island and the surrounding area.

The Garwood Ferry Bridge, built in 1933, spans the San Joaquin River as part of State Highway 4 (Charter Way). An important highway link, this bridge has not had to open for water traffic for some time.

STOCKTON

Although born on the same river as Hill's Ferry, Grayson, Firebaugh and others, Stockton was touched by the magic of the railroad and survived to become an important city. Stockton has been the major shipping point and only genuine port on the San Joaquin River from the very beginning of navigation on the waterway. In the years prior to the arrival of the railroad to the area, upwards of 1,000 steam and sailing

This post card view shows Stockton Channel as it appeared in c1905. A large paddle wheel steamer can be seen alongside the wharf. Stockton has been the head of navigation on the San Joaquin River, for most ships, since the very beginning. Photo: U.S. Archives.

vessels a year, called at the wharves along Stockton Channel and Mormon Slough. Except for the closing years of the 1850s and most of 1860, when Hill's Ferry had a legitimate claim to the distinction, Stockton has always been the head of navigation on the San Joaquin.

The arrival of the railroad brought with it many changes and a measure of prosperity, but it also signaled the beginning of the end to the first chapter of Stockton's maritime history.

When the Stockton Ship Channel was completed in 1933, at a cost of $1.5 million, Stockton became a deep water port, capable of accommodating the largest of ocean vessels. That event did not signal the return of river transportation as such, but it did turn a page in the latest chapter of Stockton's maritime history. The first deep water vessel to arrive in Stockton was the steamer *Daisy Grey,* on February 2 1933.

The channel project was largely a matter of cutting through several of the "S" turns in the river, widening it to a minimum of 100 feet and dredging the entire length to guarantee a minimum depth of twenty feet. Several new islands were created in the process, making bridges or ferries necessary in some places.

Within the City of Stockton there have been several drawbridges; most spanned Mormon Slough or other tributaries of the river.

(A.T.& S.Fe R.R.) * Mile 40.6

Just south of Rough and Ready Island, Santa Fe Railroad tracks cross the river over a bascule bridge built in 1929 by American Bridge Company. On the company's main route between Richmond and Fresno, an average of 18 freight trains and eight Amtrak San Joaquin passenger trains cross this bridge daily. One of three identical bascule bridges built for the company in 1928–29, it, like the other two, is structurally sound and still in service. In 1960, however, the company rendered this bridge permanently inoperable by welding rail across the point where it used to open.

Built in 1929, this "jack knife" type bascule span replaced a swing bridge erected here in 1913. While still well used by the railroad, this bridge is no longer required to open for water traffic.

BURNS CUTOFF

(Jacobs Road)

Below the Santa Fe bridge, the river jogs around the southeast corner of Rough and Ready Island, where it joins the Stockton Channel and the Port of Stockton.

In 1910, a steel swing span, to be known as Jacobs #1, was built across the San Joaquin River to the island as part of Jacobs Road. On the opposite side of the island, where the same road crossed Burns Cutoff, a similar structure, Jacobs #2, was built four years earlier. Both bridges were narrow, thru truss swing spans with wooden plank roadways.

In 1938, the Navy replaced Jacobs #1 with the existing swing bridge, which is located at Navy Road, 950 feet upstream from the old bridge site. With 17 feet of vertical clearance at low water, Jacobs #2, across the lightly used Burns Cutoff, rarely had to open for water traffic. It spent at least twenty years in an abandoned condition before finally being removed c1980.

(Dagget Road)

Dagget Road in Stockton runs due north from the Borden Highway and for many years served as a route to Rough and Ready Island, via a swing bridge across Burns Cutoff. Built in 1906, the bridge has not opened for water traffic in several decades. Its condition is described by the Navy as "abandoned in place mode." Standing in the open position and no doubt rusted in place, its approaches are barricaded to prevent cars from going into the water and to keep unauthorized persons off the span.

The only remaining active Navy drawbridge to the Island is the span at Navy

Looking west toward the Dagget Road Bridge in c1950. Nearly all the traffic at this site was over, not under, the bridge. While still in place, it is frozen in the open position and has not been used for at least 30 years. Photo: U. S. Coast Guard.

Completed in 1938 to replace Jacobs #1, This bridge spans the San Joaquin River at Navy Road in the Port of Stockton. Owned by the Navy, it is the only motor vehicle entrance to Rough and Ready Island.

Road, which replaced old Jacobs #1 in 1938. Although referred to as the island's "main entrance," it is actually the only vehicle entrance. The reference is no doubt a carryover from the days when the other bridges were in use.

Port of Stockton Railroad * Mile 39.7

About a thousand feet upriver from the Navy Bridge a railroad swing bridge also crosses the river to the island. Built jointly by Southern Pacific, Western pacific and Santa Fe Railroads in 1933, it is now owned by the Port of Stockton. A condition of building the bridge was an agreement whereby the city would rent the bridge at 5% of its cost per year with an option to buy the bridge within 10 years, at cost.

The only remaining drawbridge on the San Joaquin River subject to opening, it requires 12 hours notice. Definitely not a busy bridge, it opened only twice during the eleven year period 1982–1993. Both openings were for tugs with barges. While equipped with a gas engine, the bridge can also be opened by hand. Until at least 1960, its navigation lights were still oil lanterns.

This small railroad swing bridge is the only bridge on the entire San Joaquin River that is subject to opening for water traffic. As can be seen in the photo, ships do come close to it, but that's all. Its draw is much too narrow for nearly everything but tugs or pleasure craft.

FREMONT SLOUGH

The earliest known drawbridge in Stockton proper was a wooden swing bridge across Fremont Slough at El Dorado Street, about where the City Hall stands today. Built in 1862, it allowed vessels to pass through El Dorado Street to the Jacob Wagner Tannery, a short distance upstream. The upper end of Fremont Slough and Lake McLeod once formed an arm of Stockton Slough (now Stockton Deep Water Channel).

MORMON SLOUGH

Mormon Slough joins the Stockton Channel at the turning basin in Stockton, Until the beginning of the second half of this century, this slough was navigable for three or four miles above that junction. Due to industrialization in the area surrounding the slough, the realignment of highways, and the construction of freeways, the slough has all but disappeared. Its entire length is now less than one mile.

Lincoln Street * Mile 1.2

Looking west through the Lincoln Street Bridge in 1940. Seen in the near distance, down the channel, is the 1896 Santa Fe Railroad bridge at Edison Street. Just right of center, in the far background, the counterweight of the 1926 Washington Street Bridge can be seen. Photo: CALTRANS

In 1894, the city erected a narrow bridge of light construction over Mormon Slough at Lincoln Street. The swing bridge measured only 192 feet in total length, with 188 feet of that being the swing span. Its wooden roadway and sidewalk, standing about four inches above the street level, made for a very rough bump at each end of the span. No amount of work by the street department was ever able to rid the bridge of its bump for long. In spite of its flimsiness and ever present bump, it was used extensively for nearly 70 years before being retired in 1964.

Main Street * Mile 0.75

In 1903, a large wooden swing bridge was built across Mormon Slough at Main Street. Owned jointly by Southern Pacific and the City of Stockton, it carried both vehicles and trains on the same level.

Stockton's Main Street Bridge was its busiest drawbridge and also its most troublesome. Built in 1903, it had residents at their wit's end by the time it was replaced in 1927. Photo: Bancroft Library.

The scene of many bad accidents, the bridge was a constant source of complaints. Although forced to opened often for ships, and literally worn out from surface use, it managed to survive for twenty-four years. On March 4, 1927, to the delight of nearly everyone, it was finally removed. A local newspaper wrote, "...and not a single tear was shed over its demise."

A few months earlier, in late 1926, a new bridge had been completed at Washington Street, two blocks to the south.

Edison Street * Mile 1.0

This 1942 photo of the Edison Street Bridge was taken shortly after its completion. It replaced a worn out swing bridge built in 1896 by the Santa Fe Railroad. Photo: CALTRANS.

In 1896, two years after the Lincoln Street Bridge was completed, the Santa Fe Railroad erected a swing bridge at Edison Street. It was similar in appearance to the Lincoln Street span, but of sturdier construction. It served until c1941, when it was replaced by a single-leaf bascule bridge.

The new bridge, which was nearly identical to the Washington Street span, carried two lanes for vehicle traffic and a sidewalk outside of each truss. Owned by the City of Stockton, the new bridge was strictly for auto and pedestrian traffic. The Santa Fe Railroad had rerouted its tracks eastward, to a point where they no longer crossed the navigable portion of the slough.

This bridge was removed sometime before 1967. The reason for its removal is not known, but was probably related to the construction of Interstate 5 through Stockton. The slough above this point was being filled in little by little.

Washington Street * Mile 1.1

The best remembered, most needed and certainly the most looked forward to bridge in Stockton's history, was the Washington Street Bridge. Built to take the place of the ancient Main Street Bridge and to unclog a traffic bottleneck in the city's center, its opening in September of 1926 was hailed by one and all.

This large Strauss, vertical counterweight, bascule bridge was constructed by C. J. Nystead. The first highway bascule bridge in the San Joaquin Valley, it was also the first one to be operated electrically. Its 124-foot leaf was lifted by power from two 40 h.p. motors. The bridge provided a 21–foot roadway and two sidewalks. The structure cost the City of Stockton $130,000.

After serving for some 42 years, the bridge was abandoned. At the time of its removal in 1968, it had been several years since a vessel large enough to require its opening had made it that far up the slough. Even during the first twelve months of operation (1926–1927) it averaged only one opening per day. This bridge survived to become the last drawbridge over the slough.

Today, very little remains of Mormon Slough. Interstate 5 crosses what used to be part of it, but is now dry land.

Opened to traffic in September of 1926, to replace the troublesome bridge at Main Street, the Washington Street Bridge was a dream come true. Unfortunately, by the time the switch was made, a major source of complaints had nearly ended. Water traffic under the bridge had slowed considerably and openings were down to only one a day. When this photograph was taken in 1966, the bridge was no longer in use. Photo: CALTRANS.

This c1950 photo, of the Washington Street Bridge better shows the condition of Mormon Slough at this point. While the span was still able to open and close, it was seldom required to do so. Not only was there no place to go, the waterway was barely navigable. Photo: California State Archives.

SMITH CANAL

Smith Canal joins the San Joaquin River just west of Channel Point, the river's junction with the Stockton Channel. The narrow canal was originally dug to carry away sewage from a state hospital for the insane. It was also used to drain an area being reclaimed by a reclamation district. Although the waterway has never been classified as navigable by the Corps of Engineers, the State of California, through an agreement at the time the canal was dug, proclaimed it navigable.

In 1905, an iron swing bridge was constructed across the canal by what is now the Western Pacific Railroad. The bridge was retired in 1923 when replaced with the hand–me–down swing span from Lathrop. It was removed c1970 after the railroad abandoned that route.

The north side of the canal, which is now lined with attractive waterfront homes, ends about a mile above the former bridge site at Yosemite Lake. A foot bridge and a pair of freeway bridges (I–5) now cross the channel.

This Western Pacific Railroad bridge, shown here at the Smith Canal, originally spanned the San Joaquin River near lathrop. Built in 1906, it was moved here in 1921 to replace a similar span built in 1905.

Antioch * Mile 7.6

The Antioch Bridge opened one minute after midnight on New Year's Day 1926. Welcomed with a wild celebration, it was billed as "The route that connoisseurs of motordom prefer." The owners boasted that, "...unlike the ferries, the bridge is open 24 hours a day, and there is no delay, no hurrying for the ferry and it never `pulls out' just before you get there." Their brochure also invited the motorist to, "See the wonderful Delta section that produces 90% of the world's asparagus; See the Netherlands of America; Rio Vista's quaint Chinatown; Isleton's Chinatown devastated by the 'fire demon' and still in ruins."

Located 7.6 miles above the mouth of the San Joaquin River, it was the longest and most important drawbridge to ever span the river.

Built and owned by the American Toll Bridge Company, an affiliate of the Golden Gate Ferry Company, it cost the owners $1.5 million. It operated as a toll bridge under a 25-year franchise issued in June of 1923 by the State of California. By terms of the agreement, the bridge was to have become the property of the counties of Sacramento and Contra Costa in 1948. However, due to constant complaints by motorists regarding the high tolls and poor roadbed condition, the state purchased the bridge and took it over in 1940. The state purchased the Carquinez Bridge at the same time and officially added it to the state highway system on September 6 of that year.

For several years, the structure was known as the "Victory Bridge," because it connected to the Victory Highway on the north side of the span. The narrow two-lane structure joined the City of Antioch with Sherman Island. Nearly 3,600-feet long, it had a 290-foot vertical lift span located a few hundred yards from the Sherman Island shore. In spite of its 80-feet of vertical clearance at low water, it was forced to open often because of its strategic position near the entrance of the river and ship channel. Since it was also situated in an area subject to heavy fogs and treacherous tides, it was plagued with accidents for most of its existence. It was also not uncommon for the lift span to be out of commission for several days, or even weeks at a time. On one occasion, the span was closed to highway traffic for nearly six months.

On New year's Day of 1959, the S.S. *Kaimain* hit the bridge a glancing blow forcing it to close for several weeks. Damage to the structure amounted to $230,000. In 1963, the M.V *Pasadena* missed the main channel, where the bridge was raised, and tried to go under a fixed span. It made it through, but sheared off its mast in the process.

On September 4, 1970, the S.S. *Washington Bear* struck one of the bridge towers bending it almost to the breaking point. Four months and $1.4 million later it reopened to highway traffic.

This 1940 photo was taken about the time the state took over the span. The bridge, and the route often fell short of glowing promises offered by the American Toll Bridge Company and its parent company Southern Pacific. Financial trouble eventually led to its sale. Photo: CALTRANS.

During that incident, the bridgetender, Jack Davidson, was stranded, fear–stricken, high above the bridge deck when the span wouldn't come down after the accident. Afraid of heights, he refused for hours to climb down or to even move. "Every time the bridge goes up, I go up with it," he said. "And if the bridge doesn't come down, I don't either." Eighteen hours after the ordeal started, Mr. Davidson finally agreed to be strapped to a stretcher and lowered a few feet to the bucket of a snorkel truck. When it was all over, the part–time bridgetender declared, "I'm not going to work that bridge any more."

One accident, not directly attributed to a ship, occurred when a speeding car drove through a pair of barriers and plunged into the river. The span was raised for an approaching ship, but the driver was going too fast to stop. The car was recovered but no bodies were found.

The Antioch Bridge was the first really important bridge to be built in the San Francisco Bay Area. Its construction made it possible for the first time, to travel directly from Sacramento to the East Bay without the use of the ferries. In replacing the state's oldest ferry route, the bridge signaled the real "beginning of the end" of the ferries on San Francisco Bay.

It wasn't the first bridge to be planned for that location, however. In 1913, the War Department approved plans by the Oakland, Antioch and Eastern Railroad to build a vertical lift bridge across the mouth of Suisun Bay. A combination railroad and pedestrian bridge, it was to have spanned the narrow waterway between Chipps Island and Pittsburg. Plans called for a 3,000–foot steel structure with a 230–foot lift span to cost an estimated $1,250,000. The proposed bridge would have had 73 feet of vertical clearance in the closed position. It was never built.

Although the impact of the Antioch Bridge and the Carquinez Bridge (also opened in 1926) brought about the end the ferries in that area, there was an irony. By 1940, both bridges were losing so much money they were sold to the state.

The antiquated Antioch drawbridge was finally closed to traffic for the last time on November 4, 1978, when the existing high-level fixed bridge, named for Senator John A. Nejedly, was dedicated. Due to its inherent height above the river, the 1.8–mile long, concrete structure does not require a moveable span.

On February 5, three months after the opening of the new bridge and fifty three years after its own opening, the "Victory Bridge" was blown to pieces by multiple dynamite charges. The lift span, secured in the open position near the top of the towers, was destroyed by 40 well-placed charges. The towers came down in a similar manner at a later date.

In an effort to save the Benecia–Martinez Ferry from extinction, the American Toll Bridge Company was ordered to sell it to any of the cities or counties between which the ferry operated. The City of Martinez took over the ferry at one minute after midnight of December 31, 1940, fourteen years <u>to the minute</u> from the opening of the Antioch Bridge.

The toll booth of the Antioch Bridge can be seen in the center of the roadway on the lift span. The bridge toll at the time of the state takeover was 30 cents. Photo: CALTRANS.

CHAPTER 8

SAN JOAQUIN DELTA

Unlike the drawbridges across the San Joaquin River, the majority of bridges dotting its delta have had little or nothing to do with the development of any towns and have likewise been unaffected by the railroads. With but one or two exceptions, there are no towns in the San Joaquin Delta.

While marinas and resorts have developed at some bridge sites, due to the natural access to the water, most of the bridges are isolated structures with few signs of civilization in sight. The principal function of the delta drawbridges is to grant access to the islands for the purpose of agriculture and recreation.

MIDDLE RIVER

All of the bridges over Middle River were built in this century and mostly for the same reason: to help get agricultural products from the reclaimed land of the delta islands. Most are tended on a regular basis only during harvest season, and in the summer when many pleasure boats are on the waterways.

Spanning Middle River, between Union and Roberts Islands, this 1912 swing bridge served until 1969, when replaced with a fixed span. It was on Howard Road west of French Camp.

Tracy (Howard Road) * Mile 22.6

The first drawbridge to span Middle River was the Williams Bridge linking Union and Roberts Islands. Built in 1912, it saw many busy years until the late 1940s. The farthest upriver bridge, its 169–foot swing span made up nearly 90% of the structure. When replaced by a fixed bridge in 1968, it reportedly hadn't opened for water traffic in at least 15 years.

Tracy Road (J 2) * Mile 15.0

This swing span, which makes up most of the bridge's total length, was built as Jacobs #1 at Rough and Ready island. Moved here in 1948, it was installed on a newly built center pier. With water under the south side, it had only one draw. Photo: San Joaquin County Public Works.

The swing bridge that spanned Middle River at Fish Camp for many years was actually built in 1910, two years before the Williams Bridge, but not at this location. It formerly served for 28 years on the San Joaquin River between Stockton and Rough and Ready Island as Jacobs #1.

When the county decided to build a bridge at this location, it incorporated the 249-foot swing span of old Jacobs #1 into its plans. Timber approach spans and a center pier were constructed in readiness for the swing span, which was barged into position and raised onto the new pier. At this location, the bridge had water under the south side only.

Located on Tracy Road, 1–1/2 miles south of the Borden Highway, it began its tour of duty here in 1948. Built at a time when vehicles were somewhat smaller, especially trucks, its narrow (17 foot) wooden roadway originally carried two lanes of traffic. Shortly after its arrival at this location, it was made a strictly one–way bridge. Signals at each end of the bridge controlled the one-way traffic.

This bridge was replaced in 1977 by the existing fixed bridge.

Bacon Island * Mile 8.6

Until 1950, San Joaquin County operated two ferry boats between Bacon Island and Lower Jones Tract. The volume of business and the size of the loads got to the point where the ferries just couldn't handle all the traffic. After much agonizing, the San Joaquin County Board of Supervisors, in 1949, authorized construction of a drawbridge to replace the ferries. Fortunately for the taxpayers, someone was looking out for their interests. It was discovered that Sacramento County had a spare swing span for sale.

A 335-foot thru truss swing span, built in 1905, it was originally part of the first Steamboat Slough Bridge at Grand Island. It was used there until 1937, when it was moved to Three-mile Slough on highway 24, south of Rio Vista. When a new vertical lift bridge was built at Three-Mile Slough in 1949, the swing span became surplus.

The state's most traveled swing span is shown coming around the bend heading for its final location at Bacon Island. Moved here in 1950, it previously spent 32 years at Grand Island and 12 years at Three Mile Slough. Photo: San Joaquin County Library. The photo below shows the bridge at Bacon Island shortly before being replaced in 1995 by the existing swing bridge.

San Joaquin County bought the span and, with the aid of two 100-ton floating cranes and a barge, brought it to its new location. The 165-ton structure was lifted off the barge and placed onto a newly constructed center pier. Total cost of the bridge at its new location came to $161,891. The swing span and its transportation cost, plus the center, pier came to $55,891. The remaining $106,000 was for 750 feet of timber trestle approaches and rest piers. Since the cost of a new moveable span alone, could have run as much as the total project, the saving was substantial.

Shown in the open position, during the last stages of construction, is the new Bacon Island Bridge. Costing the County of San Joaquin $7,000,000, it is expected to serve for a very long time and be money well spent.

Not unlike the bridge at Tracy Road, the center pier was located adjacent to the shore, leaving only one navigable channel. It was tended eight hours every day (9 a.m. - 5 p.m.) during the summer months and the same hours, Thursday through Monday only, during the winter. Twelve hours notice required at all other times. It was not uncommon for the bridge to be open longer hours, however, especially in the summer when water traffic was heavy.

Narrow, and seemingly held together by the many steel patches in the wooden roadway, the span was limited to signal-controlled one-way traffic. To the drivers of the many large trucks that used this bridge daily, each crossing must have felt like it could be the last.

In late summer of 1995, San Joaquin County completed a concrete and steel bridge a hundred or so yards downstream from the old bridge. The 760-foot long, $7,000,000 structure, with a 335-foot swing span, carries two, wide traffic lanes. This, California's newest drawbridge, is most welcome.

At the time of this writing, the old bridge was still in place and for sale. According to a CALTRANS spokesperson, it is unlikely this span will live to see another tour of duty somewhere else.

(State Route 4) Victoria Island * Mile 15.1

One of the more attractive bridges spanning the river is Middle River Bridge on the Borden Highway, at Victoria Island. With a thru truss swing span and pony truss approach spans, the 547 foot bridge presents a pleasing profile.

One of the three drawbridges built in 1915, it is a link in the highway that skirts the southern perimeter of the delta joining Stockton with the East Bay, via Antioch and Martinez. Now exempt from opening for water traffic, it is, none the less, a well built and maintained bridge. One of its last operators suggested, "It could survive to become the oldest moveable span in the state.

A small and popular marina and restaurant is located near the bridge site.

Spanning Middle River at Victoria Island, this 80-year old bridge does not look its age. Although exempt from opening, it is host to plenty of highway traffic.

(Santa Fe) Middle River Station * Mile 9.8

On October 28, 1898, the San Francisco and San Joaquin Railroad, which later became the Santa Fe Railroad, completed a swing bridge over the river at Noorland (Middle River Station). Located on their Oakland/Stockton route, the tracks run east and west over a 1.5 mile wooden trestle, separating Bacon and Woodward Islands.

In 1928–29, the American Bridge Company fabricated three steel, plate girder bascule bridges for Santa Fe. One was destined for the San Joaquin River, one for Old River and this one across Middle River. This one began service in June of 1930. While still in place, it has not opened for water traffic for at least five years, perhaps much longer. Twelve hours notice is required.

Looking a little like a praying mantis, this plate girder bascule bridge belongs to the Santa Fe Railroad. One of three look-a-likes built for the company in 1928-29, it was erected here in 1930. Just out of sight to the left is the Orwood Bridge. Still in use by the railroad, this bridge seldom has to open.

OLD RIVER

Old River leaves the San Joaquin south of Stockton, threads its way through the delta and then rejoins the San Joaquin again near Frank's Tract.

(State Route 4) Victoria Island * Mile 14.8

Another of the aforementioned drawbridges built along the Borden Highway in 1915, this bridge crosses Old River about six miles northwest of Byron at the San Joaquin and Contra Costa County line.

Rebuilt in 1932, the 256–foot swing span of this substantial bridge gets plenty of action. Tended 16 hours a day in the summer and eight in the winter, four hours notice is required at all other times.

Old River Bridge at Victoria Island. Island. The most westerly of the three bridges built on the Borden Highway (State Route 4) in 1915.

(Santa Fe) Orwood * Mile 10.4

At the western end of the wooden trestle separating Orwood and Middle River Station, the San Francisco and San Joaquin Railroad, now the Santa Fe Railroad, completed another swing bridge on March 6, 1898.

That bridge was replaced by the existing bascule bridge on May 19, 1929. One of three such bridges built by the American Bridge Company for Santa Fe, it was the first to be installed. In 1926, a year for which records are available, more than 1700 vessels passed through the draw of the old swing span. Although the type of craft using the waterway has changed from commercial to mostly recreational, the route is still used extensively. This is one of only a handful of drawbridges in the state that is tended 24 hours every day.

To prevent interruption in train service during construction, trains were detoured over a temporary bascule bridge built alongside the swing bridge. After the old structure was removed the present bascule bridge was erected on the site. The same method was reportedly used at all three locations.

The swing bridge, with a 202–foot swing span, provided two draws of 75-feet each. The existing bascule bridge provides one clear 90-foot channel. Although electrically operated, a 1926 Buda diesel engine is on hand as a standby.

GRANT LINE CANAL

(Tracy Boulevard) * Mile 5.5

The only double-leaf bascule bridge ever built in the San Joaquin valley is this concrete and steel structure over the Grant Line Canal. Dedicated on November 5, 1959, it cost the County of San Joaquin a half-million dollars. It replaced a swing bridge that is now serving at Connection Slough.

Between 1882 and 1890, Thomas Williams and Daniel Bixler built the Grant Line Canal that runs east and west between two points on Old River near its junction with the San Joaquin River. Built north of the "Grant Line," on land granted to them for the purpose of reclaiming the swamp and overflow lands, the canal is situated between Middle and old River. The federal government also obliged them by building a dam at Paradise Cut, thus greatly aiding the draining of the land.

At a later date, the Fabian and Bell Canal was dredged parallel and south of the Grant Line Canal. Together they formed a barrier to trade between the rich, productive land to the north and the shipping point at Tracy, to the south.

In 1904, the San Joaquin County Supervisors voted to construct a bridge across the canals and bids for the job were taken. The contract was awarded to Clark and Kerney Company, for the low bid of $18,770. Bixler conveyed land to the county for a road across Union Island and for the north bridge approach. Fabian in 1906, conveyed two acres of land for the southern bridge approach.

The structure, when completed, consisted of a 230-foot thru truss swing span over the Grant Line Canal and a 126-foot through truss fixed span over the Fabian Canal. The total length of the bridge, including approach spans, was 496 feet. With the acceptance of the bridge by the Board of Supervisors on May 8, 1906, the new route through the delta between Stockton and Tracy was opened and the barrier removed.

Time and the advent of bigger and wider trucks/trailers brought about severe load restrictions necessary for the safety of the bridge and its users. Once again, a man made barrier was standing in the way of the efficient transportation of the delta's agricultural production. Also once again, a farsighted County Board of Supervisors was up to the challenge and approved the construction of a new, bigger and better bridge to replace the 53–year old structure.

The new $588,847 bridge, built by Erickson, Phillips and Weinberg, was dedicated on November 5, 1959. Approaches added another $95,000, for a total cost of nearly $700,000. The new structure consists of a double leaf bascule girder span and twelve reinforced concrete slab spans. It provides a 28-foot clear roadway for its 471-foot length and features lightweight concrete decks on the two 61-foot bascule leaves. There are no undue restrictions on this bridge.

This bridge is located on Tracy Boulevard (J2), 6–1/2 miles north of I–205. Twelve hours notice is required for opening.

CONNECTION SLOUGH

Mandeville, Bacon Island * Mile 2.5

Delta Farms Reclamation District 2027 completed a pontoon bridge across the slough on September 1, 1947. The bridge consisted of three 87–foot floating sections, one of which could be moved aside to open a channel for passing boats. The bridge replaced a free ferry operated by San Joaquin County at a cost to the taxpayers of $13,000 a year.

Built in 1906 at Grant Line Canal, the swing span, which makes up most of this bridge, was moved here in 1966. At 89 years of age, it is still active, but has old age problems.

In 1966, the pontoon bridge was replaced by a steel bridge that included the 230–foot, 66-year old swing span from the original Grant Line Canal Bridge. Now 89 years old (1995), it is still in service and still operated by the Reclamation District. The pontoon bridge provided zero vertical clearance when closed, but with only 2 or 3 feet of clearance at high water, the swing bridge is not much better. W with so little vertical clearance, this span is required to open several times a day during the summer, especially on busy weekends. Tended 16 hours daily in summer and 8 hours in the winter, four hours notice is required at all other times.

TURNER CUT

Delta Farms Reclamation District 2030 * Mile 2.3

For nearly twenty years before the existing drawbridge was completed here, the reclamation district had tried in vain to get permission to build a fixed bridge across the cut. Because the waterway was, and still is, used extensively for all manner of boating activities, resistance to the idea was great.

During the years of petitioning for a permit, hundreds of letters were written to the county, state and U.S.Coast Guard protesting a fixed bridge over the waterway. Nearly every boating organization and marina in the northern half of the state voiced their objections.

The reclamation district wanted better access to and from their land, but the cost of a fixed bridge with enough vertical clearance to satisfy everyone seemed prohibitive. Finally in 1979, plans were presented for a proposed moveable bridge across the cut that was eventually accepted.

After nearly six years, from drawing board to completion, the bridge became a reality in 1984. Opened on December 20, with a free BBQ lunch provided by the trustees of the reclamation district, the $800,000 bridge was dedicated as the "Zuckerman Brothers Bridge."

The Zuckerman Brothers Bridge over Turner Cut. This retractable bridge is the only one of its kind in California. Built in 1984, at a cost of $800,000, it is owned by Delta Farms Reclamation District 2030.

Maurice and Roscoe Zuckerman were pioneer farmers in the San Joaquin Delta. Together they farmed thousands of acres on McDonald Island, Mandeville Island and the Rindge Tract. The Zuckerman Brothers contributed mightily to the growth of California agriculture and are part of the delta's history. They were key participants in the early history of delta farming and the reclaiming and safeguarding of the delta islands. Maurice Zuckerman was succeeded in farming by his son, John, and Roscoe, by his son, Alfred.

The only structure of its kind and size in California, it consists of a 35–foot–long moveable span that rests on two fixed spans. The moveable span is operated hydraulically and moves horizontally across the channel. Total length of the structure is 470 feet.

Tended 24 hours a day, this bridge is posted for one–way traffic only. The bridge was designed and built so the moveable span could be operated electronically, via key control boxes on the approach spans. Due to the amount of traffic on and under this bridge, however, it was decided to have a tender on the span at all times. The bridge can be opened or closed in about five minutes.

The navigation channel is 25 feet between fenders; vertical clearances range from 16 to 19.5 feet in the closed position. It opens for water traffic about a half dozen times a month in winter and up to twenty times a day on a busy summer weekend . Its generous clearance in the closed position no doubt keeps the number of openings down.

BISHOP CUT

Eight–Mile Road * Mile 1.0

In 1927 a steel swing bridge was constructed over Bishop Cut on Eight–Mile road. North of Stockton and 8-1/2 miles west of Highway 99, this thru truss bridge linked King Island to the "mainland."

The narrow, 322-foot long bridge had a 191-foot, electrically operated, thru truss swing span. Originally marked for two lanes, its 18-foot roadway was later changed to one-way traffic controlled by traffic signals.

In 1989, a concrete and steel swing bridge was built to replace the 62-year old structure. Carrying two-way traffic once again, its short swing span provides two narrow draws for water craft.

In an effort to make it accident-free, crossing arms, traffic signals, flashing lights and a set of horns, all warn highway traffic when the bridge is open. "Drawbridge ahead" signs also warn motorists of its existence as they approach the bridge. Located in the "middle of nowhere" and looking nothing at all like a drawbridge, all of the above precautions are probably warranted. Twelve hours notice is required to open.

The new Bishop Cut Bridge is one the state's newest draw-bridges, but looks nothing at all like the one it replaced. Extra warning devices are necessary to warn motorists of its presence. Photo: CALTRANS.

HONKER CUT

Eight–Mile Road * Mile 0.3

Two and 1/4 miles farther west on the same road, another swing bridge spans Honker Cut, between King and Empire Islands. Dedicated on October 28, 1936 as the "Empire Drawbridge," it too, in effect, connects Empire Island to the "mainland." Although slightly longer than the former Bishop Cut bridge and without a control house, to the casual observer, they appeared to be like two peas in a pod. Manually operated, it too requires 12 hours notice.

Both structures replaced ferry services and can take credit for much of the extensive agricultural development of these islands.

Once powered by an ancient gasoline engine, the 204-foot swing span of this bridge is now manually operated. Fortunately for the bridge tender, it isn't required to open often.

CHAPTER 9

HUMBOLDT BAY TRIBUTARIES

Humboldt Bay, the first important port north of San Francisco, has been a lumber port since before the turn of the century. Consisting of two shallow basins (North Bay and South Bay), served by dredged channels, it contains large marshy areas fed by many sloughs.

Eureka, on the North Bay, four miles north of the harbor entrance, is the area's largest city. Arcata, northwest of Eureka, is next in importance as a shipping point. Samoa, on the western shore opposite Eureka, was once an important railroad terminal.

There are no important towns in the South Bay, but at one time it was the site of a sizeable ship–building industry. Many of the wooden steam schooners that plied the coast early in the century were built at Fairhaven, a place no longer on the map.

Although far removed from the scene of most of the of state's drawbridges, the sloughs of Humboldt Bay have been host to at least ten. Unfortunately, few photographs are available and today only one drawbridge remains.

EUREKA SLOUGH

A tidal stream, Eureka Slough extends northeasterly from the main channel above Eureka for about three miles. Its width and depth vary greatly according to distance from its mouth.

In 1931, four drawbridges spanned this waterway: a railroad swing bridge 2,300 feet above the mouth; a highway vertical lift bridge 200 feet above the railroad span; a railroad bascule bridge at mile 1.75; and a highway bascule bridge at mile 2.25.

Most of the traffic on the slough around the turn of the century was made up of small lighters and barges drawing as little as two feet of water. Outgoing cargo consisted of forest and farm products. Upriver cargoes were general merchandise and fuel. Logs were also towed down river on the tide.

Today there is little or no commercial use of the waterway, especially above the only remaining drawbridge.

(North Coast R.R.) Eureka * Mile 0.5

The earliest drawbridge to span Eureka Slough was a wooden bascule bridge built before the turn of the century by the North Western Pacific Railroad. That bridge was replaced with a steel swing bridge in 1917 by the Eureka Southern Railroad. Boasting an electrically operated 188–foot swing span, their new bridge opened to rail traffic on November 16, 1917. According to the company's records, the span's busiest years were before 1930. During its 58 years of service it averaged less than four openings a year and between 1948 and 1953 it did not open at all.

The swing bridge in the foreground was built in 1916 and belonged to the Northwestern Pacific Railroad at the time of this photo. The fixed bridge to the left is the existing Highway 101 bridge. The closeness of the two spans and the angle of the railroad bridge was a problem in earlier years when traffic on the slough was a factor. The swing bridge was removed in 1975. Photo: CALTRANS.

During the 1960s, the bridges' owners (Northwestern Pacific Railroad) tried repeatedly to get approval to make it a fixed bridge, but with no success. Finally, in 1972, they gave up and agreed to build a short–lift vertical lift bridge.

Completed in 1975, the new bridge has a 100–foot lift span that provides a 67-foot channel. The term, short–lift referes to the height the span has to lift. In this case, it is only high enough to provide the same vertical clearance as the highway fixed bridges upstream. The railroad was directed to use as much of the old bridge in the construction as possible and then remove all trace of the rest of it within 90 days.

This unique bridge over Eureka Slough is the only remaining drawbridge in "Northern California." Owned by the North Coast Railroad, it is currently exempt from opening. Photo: U.S. Coast Guard.

Like no other bridge ever built in California, it has no cables, pulleys or counterweights. its moveable span is raised and lowered hydraulically by a diesel powered compressor.

The only existing drawbridge on the slough, and in fact, in Northern California, it is no longer required to open for water craft. It is, however, subject to orders from the Coast Guard that could force the owners to put it into operating condition within 60 days of notification.

In 1984, the railroad line over this bridge was abandoned by the North–Western Pacific Railroad. The following year, the Eureka and Southern Railroad took over the line, but went bankrupt in 1987. The present owner of the railroad and the bridge is the North Coast Railroad.

(US 101) Eureka * Mile 0.55

The best known drawbridge across the slough and in the entire Humboldt Bay area, for that matter, was the bridge on Highway 101 between Eureka and Arcata. A wooden vertical lift bridge with a 75-foot lift span, it carried a narrow 21-foot roadway. Located just yards away from the railroad bridge and built by the state at a cost of $84,000, it first opened to traffic on March 1, 1920.

Replacing a worn out swing bridge built many years earlier, this bridge survived until October 9, 1942, when replaced by a fixed concrete bridge.

The closeness and alignment of the railroad and highway drawbridges was a constant problem, especially during strong winds. Both bridges crossed the slough at an angle and the highway span was on a bend of the waterway. Getting a tug with a barge or 80–foot log boom through the two draws called for some delicate handling. Accidents, some of which closed one or both of the bridges for extended periods of time, were not uncommon.

The last highway bridge to span a waterway in the Eureka area was this wooden structure over Eureka Slough. From 1920 until 1942, it carried the north and south traffic of Highway 101. Photo: CALTRANS.

(Pacific Lumber Company) Eureka * Mile 2.0

The Pacific Lumber Company built a railroad bascule bridge across the slough in 1921. Situated two miles from the mouth of the slough and built as late as it was, it seldom had to open. The date of its removal is unknown.

Eureka * Mile 2.6

The furthest upstream bridge was a small, hand operated, wooden bascule bridge built c1928 and owned, in 1941, by a Mr. N.M. Derby. There is no record of how often, if ever, it opened for water traffic or when it was removed.

MAD RIVER SLOUGH

During the early years of the century there were wharves on Mad River Slough where seagoing lumber vessels loaded logs and other forest products. Logging in that area slowed considerably during the 1920s, however, and use of this slough dropped off accordingly.

Although the Corps of Engineers reaffirmed the navigability of the waterway in 1931, by 1933 there was so little water traffic on the slough, it was decided moveable bridges were no longer required. Not only had some of the spans never opened for passing boats, at least two of them were completely inoperable. Owners of all four bridges in place at that time made application to remove or convert their moveable spans to fixed spans. All applications were approved and before the end of 1935, fixed bridges were in place at all four locations.

(N.W.P.R.R.) Arcata * Mile 0.25

Mad River Slough was bridged in 1898 and again in 1913, by the Northwestern Pacific Railroad. Both structures were small, manually operated wooden bascule bridges across the mouth of the slough. The company's last moveable bridge at that location was built in 1927 and removed in 1933. It reportedly had never opened for water traffic.

(State Route 255) Arcata * Mile 0.5

Humboldt County erected a bascule bridge (35–foot draw) across the slough adjacent to the above railroad bridge in c1917. An application to make it a fixed bridge was approved and the project was completed on December 13, 1932. The span had been inoperable for years and no one could remember ever seeing the bridge open for any reason.

(Humboldt Northern R.R.) Arcata * Mile 1.2

A small wooden bascule bridge (35–foot draw) was constructed across the slough by the Eureka and Klamath River Railroad in 1897. While probably not unique, it was a bit unusual. The moveable span was raised by a counterweight in the form of a small four wheel car on tracks down the back side of the king post. To open the bridge, the car was allowed to descend the tracks.

That bridge was replaced by a fixed bridge on August 10, 1934. It is believed to have been the original structure there.

(Humboldt County) Arcata * Mile 2.5

On February 23, 1918, the county completed a small bascule bridge (38–foot draw) across this waterway near the head of navigation. It was removed at the county's request on February 17, 1932. At the time of its removal, the hand winch was found to be rusted and broken and the lift tower in a state of near collapse. A farmer living near the bridge claimed that although it had never opened, he once saw a small, take-down pile driver on a barge under tow, pass <u>under</u> the bridge during a very low tide.

FRESHWATER SLOUGH

An arm of Eureka Slough, this tidal waterway saw some active years near the beginning of the century. Small launches and barges under tow hauled oil, logs and general merchandise to Humboldt Bay. In 1931 three bascule bridges, all dating from the 1920s, spanned the slough. It is unlikely they were the original bridges at those locations, however. Today there are no drawbridges over Freshwater Slough.

Eureka * Mile 0.5

A wooden bascule bridge was erected across Freshwater Slough by Peter H. Christianson in c1900. It survived until April 25, 1929, when a new bascule bridge was completed at that same location. The new bridge was a simple affair, with a 36–foot wooden truss span raised with the aid of a hand winch.

When converted to a fixed structure in April of 1974, it was non-operational and no one remembered ever seeing it open. The property is now owned by Dutra Farms.

Eureka * Mile 1.6

A mile upstream, another bascule bridge, listed as being on a private road and owned by M. Brazil, spanned the slough from the early 1920s to about 1930. It provided a 30-foot channel over a seldom used waterway.

Eureka * Mile 1.85

The furthest upstream bridge was located on a private road listed as a cattle crossing. Also a bascule bridge, it provided a 28 foot channel. Built c1930, this wooden trestle bridge was replaced with a concrete fixed span in 1937. The drawbridge, which was by then inoperable, had not opened in years and the trestle-was in danger of collapsing. With the head of navigation less than a half mile above the bridge and no commerce on the waterway, the Corps of Engineers agreed a moveable bridge was no longer necessary. The route was the only connection to several dairy farms in the area.

ELK RIVER SLOUGH

Eureka * Mile 0.5

A Railroad swing bridge was erected across Elk River Slough in 1886. According to available records, it opened fewer than five times during the ten years it was in place. After a fixed bridge was built downstream it was exempted from opening. The name of the railroad is not recorded.

It is believed that some of the early, short–lived logging railroads bridged all four of the above sloughs with both fixed and moveable spans. Nautical charts for the period 1900–1930 show some of the crossings, but do not always identified them as to type (fixed or moveable).

CHAPTER 10

SOUTHERN CALIFORNIA

LONG BEACH INNER HARBOR ENTRANCE

Los Angeles & Salt Lake R.R.

In 1891, the San Pedro, Los Angeles and Salt Lake Railroad built a trestle bridge across the waterway that separated present–day Long Beach from Terminal Island (formerly Rattlesnake Island). Located at approximately the same spot where today the Gerald Desmond Bridge stands, it became the first rail link between Long Beach and Terminal Island.

In 1908, the same company, under its new name, Los Angeles and Salt Lake Railroad and by then, partly owned by the Union Pacific Railroad, began construction on a bascule bridge to replace the trestle bridge. The span was completed and opened to rail traffic on February 26, 1909. During the construction of the bridge, the tracks were carried over a temporary span.

Plans for a turning basin and deep water port above the bridge were already under way when the trestle bridge was constructed in 1891. At that time, the railroad's owners were warned that the day would come when any bridge across the waterway would have to include a moveable span. In 1906, when the waterway was declared navigable, the War Department called in its chips by ordering the railroad to remove its "obstruction" to navigation.

Although boat owners had been complaining for years about the trestle across the channel, the railroad was in no hurry to abide by the government's order. They used every delaying tactic available to avoid complying with the order. They made a serious mistake, however, when they removed a section of the bridge to let a dredge through. The government pounced on them and saw to it that the section was left out, except to allow trains to cross. It was not long before construction on the bascule bridge began.

Erected by a local firm, Llewellyn Iron Works, the Scherzer heel trunnion bascule bridge carrying a single set of tracks cost the railroad $275,000. Its 180–foot leaf, one of the largest of its kind in existence, was raised easily and quickly by two 20 h.p. motors. In an emergency it could also be opened manually. Electricity was provided by the Pacific Electric Railway Company, whose terminal was located in Long Beach, a short distance from the bridge.

The first ship to pass through the new bridge was the S.S. *Hubbard* in late 1910. The harbor was not officially opened to ocean going ships until 1911, however. Water traffic on the channel increased dramatically year by year, as did the number of bridge openings and mishaps. The Los Angeles and Salt Lake Railroad was taken over completely by the Union Pacific in 1921. By 1935 the railroad had abandoned the route and removed the bridge.

During the years between the demise of this structure and the construction of the Navy's pontoon bridge in 1944, passengers from Long Beach were carried to and from the island by water taxis.

After operating the bridge for 18 of its busiest years, S.E.Fulton, when asked if there was anything glamorous or romantic about his job, said, "No, there's no romance or adventure in this job. When they whistle for passage, I raise the bridge. When they pass through I lower it. That's all."

The S. S. *General Hubbard* is shown passing through the draw of the Los Angeles & Salt Lake R.R. Bridge in 1910. It was the first ship to go through this, Southern California's first drawbridge. Photo: Long Beach Library.

(U.S. Navy) Pontoon Bridge * Mile 7.O

In 1944, due to increased activity at the Navy's Roosevelt Base on Terminal Island, the Navy needed an additional auto route to the island. Using approximately the same location formerly occupied by the Los Angeles and Salt Lake Railroad Bridge, the Navy constructed a highway drawbridge across the entrance channel to Long Beach Harbor.

Although a tunnel under the channel had been under consideration, the urgency of the situation caused the Navy to decide on a pontoon bridge. Completed and opened to traffic in October 1944, it was patterned after a retractable pontoon bridge on Seattle's Lake Washington. Intended as a temporary structure, it was scheduled to be removed six months after the end of the national emergency.

Built by Guy F. Atkinson Company, it cost the Navy $669,000. The two 51 X 135 X 13–foot barge–type pontoons were constructed of reinforced concrete to save critical steel during the war.

Locked together at the center and secured to ramps from the approaches on each end, the two pontoons formed a four–lane bridge across the 180–foot waterway.

"When the bridge is opening the two pontoons are stabilized by a double set of outlying abutment towers, which also serve to partially support the ramps and a fabricated steel yoke. The purpose of the yoke is to guide and control the movement of the pontoons, which freely retract through the yoke and under the

This aerial view shows the Navy's retractable pontoon bridge in the open position. To the left is the nearly completed Gerald Desmond Bridge. The largest pontoon bridge ever to span a California waterway, it served for 24 tumultuous years. The view is to the east. Photo: Port of Long Beach.

ramp to clear the ship channel. The lower end of the ramp and yoke are supported on rollers in contact with the traffic deck; the upper ends are pivoted to the abutment towers and approach. In this way both ramp and yoke are free to follow the vertical tidal movements of the pontoon.

Each 120–foot ramp is divided into equal sections by a cross beam support. The upper section is semi–stationary, being fixed to the approach above and hinged to the mid-beam support below. The lower section is hinged to the mid–beam support with its free end resting on the pontoon deck. Adjustments can be made to equalize the grade on the two sections." *Western Construction News* , July 1945.

The bridge was designed to operate within the maximum tidal range (about 10 feet) with little room to spare in its articulated joints. The ramps were hinged and provided with winches to adjust the slope of the ramps as the tide changed.

Over the years, the land and the bridge subsided at the rate of one foot per year, due to the extraction of oil. Before the subsidence was stopped (sometime in the 1960s, I seem to recall), the entire area where the bridge was located had subsided over 20 feet. As the bridge subsided beyond its operating limits, the pontoons were retracted back under the ramps, closing the bridge to vehicle traffic for ever increasing hours each day until the tide fell low enough for the bridge to operate again.

In 1955, in an effort to remedy the situation on the heavily–traveled Ocean Boulevard, the bridge was rebuilt to a higher elevation, 100 yards to the south. It opened to traffic at its new location on April 28, 1955. During the eleven day transition period, the public had to use the Commodore Heim Bridge.

The subsidence problem did not end with the move, and within a few years the bridge was again two–blocked for many hours a day. During the first few years of operation an average of 20,000 + cars a day passed over the span and some 35 ships per day passed through its draw. *Ray Berbower, Assistant Chief Harbor Engineer (retired) for the Port of Long Beach.*

Because of the higher elevation at the new site (11 1/2 feet), a passageway could be provided on the shore side of each abutment tower to allow small craft to pass through without having to open the bridge. The two 30–foot channels had up to 18–feet of vertical clearance. Unfortunately, the increase in elevation sharply increased the angle of the ramps at low tide. Busses sometimes scraped bottom and more than one big–rig truck and trailer had to be towed up the ramps. Auto traffic had to endure a wild roller–coaster type crossing during bad weather and, long delays during periods of high tides. The roller coaster effect is evident in the final chase scene of the movie '*It's a Mad, Mad, Mad, Mad World,*' in which the bridge is featured. In spite of its numerous faults, the bridge saved many commuters as many as 20 miles a day. During its early years, it was opened to vehicles only during commute hours.

Like most drawbridges, especially busy ones such as this, it had its share of accidents. On three occasions, vehicles managed to drive through the warning gates or over the guard rails and into the channel, twice with loss of life.

In 1948, operation of the bridge was turned over to the City of Long Beach. Two years later, the Navy presented the bridge to the city along with $22,000 to help pay the cost of its eventual removal.

Although the bridge opened for the last time on July 1,1968, it remained in place for several months. On August 28, 1968, Guy F. Atkinson, the same contractor that built the bridge, was awarded the contract to remove it. The company's bid of $64,000 won because they accepted the two pontoons as value received. The bridge remained in the open position until removed later in the year.

The many problems of the pontoon bridge and the increasing size and number of ships using the waterway, dictated the need for the Gerald Desmond Bridge, which was completed on July 30, 1968.

WEST BASIN

(So.Pacific.) * Mile 3.0

In 1910, Southern Pacific, wishing to further develop the West Basin, asked the government for permission to replace their aging trestle bridge. The U. S. Corps of Engineers agreed, on the condition that any replacement must contain a moveable span if the tidelands beyond it were to be dredged. The trestle was removed and a bascule bridge was built over its bones.

Built across the mouth of West Basin in 1911, it was the first Strauss bascule bridge in the Los Angeles area. Designed by the Strauss Bridge Company of Chicago and built by American Bridge

Company, it carried separate tracks for Southern Pacific trains and Pacific Electric Interurban cars. The Interurban carried passengers between Los Angeles, Wilmington and San Pedro. The tracks were previously carried across the tidelands via a fixed trestle bridge.

By the 1930s, the bridge was a monumental bottleneck for rail commuters. With 70 or more openings a day, rail traffic was often delayed for two or more hours daily. All attempts to have the bridge exempted from opening during certain hours failed. Operated as a municipal joint venture by the Belt Line Railroad, the bridge was often an embarrassment to the city.

With water traffic multiplying annually, accidents causing damage to some part of the bridge were commonplace. Even the 187–foot bascule leaf was struck on more than one occasion. *The News Pilot,* on June 13, 1930, predicted its eventual removal and claimed it was a mistake from the beginning. "Its removal, when it finally comes, will be an important step in the advancement of the harbor and also an achievement in the drive for better San Pedro transportation."

During WW II, the Coast Guard closed the bridge a time or two, for fear it might be blown up by saboteurs and end up blocking the harbor.

Amazingly, in spite of all its problems and controversy, this bridge managed to survive for 47 years. As a parting shot, however, a passing freighter hit and damaged the bridge about a week before it was to be removed.

This Strauss bascule bridge spanned the mouth of West basin from 1910 until 1957. Owned and used by Southern Pacific, it also carried cars of the Pacific Electric Interurban Railway. Photo: Historical Society of Long Beach.

CERRITOS CHANNEL

(Union Pacific.) * Mile 5.2

Sometime before 1921, The Union Pacific Railroad built a bridge across Cerritos Channel between Wilmington and Terminal Island. A wooden trestle bridge, it contained a swing span and carried a single set of tracks. Navigation was through the north draw only, which provided a clear 70–foot channel. Shortly after the completion of the Badger Avenue Bridge in 1924, this, the Channel's first drawbridge was removed.

U. S. Navy (Temporary) * Mile 4.4

On July 16, 1941, the Navy began construction on a "temporary vehicle access bridge" across the channel to provide another route to Terminal Island. At that time, the Badger Avenue bridge was the only direct vehicle route to the Navy's installation. With the threat of war and a great increase in activity on the base, there was much concern. The Commodore Heim Bridge was on the drawing board, but it was still a few years away.

Opened to traffic in 1942, the new bridge consisted of a long wooden trestle with a manually operated swing span near the center. It carried two traffic lanes and two pedestrian walks and provided an 18-foot navigation channel under the north side only. Erected parallel to the Badger Avenue Bridge, with barely enough room to swing open, it connected to Ford Avenue, adjacent to the Ford Motor Plant.

After serving for fewer than four years, removal of the swing span and center pier began in May of 1946. The trestle spans were used during the early stages of the construction of the Heim Bridge, then removed.

This 1942 sketch-like photo is of the Navy's temporary swing bridge at Ford Avenue. A small section of the Badger Avenue Bridge can be seen in the lower left hand corner. The large building in the background is the Ford plant. Photo: Long Beach Library.

Badger Avenue * Mile 4.4

The Henry Ford (Badger Avenue) Bridge was designed in 1922 by the Strauss Bascule Bridge Company. Fabricated in 1924 by the American Bridge Company, it was erected by Ross Construction Company for the Port of Los Angeles. Situated upstream from the main Los Angeles Harbor entrance, the bridge connects Wilmington with the north side of Terminal Island. It was used by trains of the Union Pacific Railroad and the Harbor Belt Lines.

Located on Ford Avenue (formerly Badger Avenue), it was named the Henry Ford Bridge as a concession to the Ford Motor Company for building their plant there. The plant was demolished years ago and the land purchased by the Port of Long Beach. The bridge is built on a perpetual easement from the City of Long Beach.

Taken from the west, this 1994 photo shows the Badger Avenue Bridge in the open position, with the Commodore Heim Bridge in the background. Construction will start in 1996 on a vertical lift bridge to replace the bascule span. Photo: U. S. Coast Guard.

A low–level, double leaf, triple truss trunnion bridge, it has two 110–foot leaves providing 180 feet of horizontal clearance. Two 50–foot tower spans and two 220–foot wooden trestles are located at each end of the moveable spans. At the time of its construction, this $980,000 bridge was one of the largest and heaviest of its kind ever built. Two electric motors operate a rack and pinion system that powers the individual leaves.

The bridge was originally equipped with two sets of railroad tracks, two vehicle lanes and one sidewalk. After WW II one set of tracks was removed.

Due to excessive pumping of oil in the area, the bridge and its approaches subsided five and one–half feet during the period from July 1940 to June 1950. Fourteen hydraulic jacks, each with a 350–ton capacity, were required to raise the bridge and approaches seven and one–half feet. The extra two feet were to (hopefully) take care of any future settlement. Unfortunately, it was not enough. In 1957 the bridge had to be raised another seven feet. Since 1957, the oil companies have been injecting water into the underground strata which has successfully stopped the subsidence. The bridge was closed to traffic for more than two years during the lifting process.

Shortly after the completion of the Commodore Schuyler Heim Bridge, all vehicle traffic was taken off the Badger bridge and diverted to the new structure.

Although owned and maintained by the Port of Los Angeles, all railroad improvements are paid for by the Union Pacific Railroad. Located in the City of Long Beach, the bridge is the only rail link to Terminal Island. Until a few years ago, the railroad served the Port of Los Angeles, Port of Long Beach, the Long Beach Naval Station and shipyard..

Thought to be a key element in the future development of Terminal Island, the bridge recently underwent extensive refitting. Just a few of the changes included the addition of a second set of tracks on the west side of the bridge; replacement of the wooden approaches with concrete spans; replacing the bridge's electrical system; and sandblasting and painting the structure.

In 1985, the bridge was designated Number 53C–61 on the National Register of Historic Places and given the status rating of three. The historic significance of this bridge is due to its being one of the longest heel trunnion bridges ever built, being the only bascule bridge in California south of San Francisco, and its impact on the development of the port and surrounding communities.

Due to the infrequency of rail traffic over the bridge and the increase in the number of openings each day, the bridge has been maintained in open position for the past several years.

On June 20, 1994 it was announced that this bridge is to be replaced by a lift bridge similar to the Heim bridge. In December 1994, it opened and closed ceremonially for the last time before removal. Actual construction on the new bridge is scheduled to start in 1996.

The old bridge has been offered to the highest bidder, but will be given away if necessary. Even for free, however, it wouldn't be a bargain. Due to its designation as an Historical Place, it will have to be carefully dismantled, an expensive process. If no one wants the bridge and the city ends up removing it, they plan to save the control house and put it on display somewhere in the harbor area.

The Badger Avenue Bridge has been seen in dozens of Hollywood movies and many celebrities have crossed it. One such celebrity was Al Capone. He spent time in the Terminal Island Federal Prison when it was located there.

Commodore Schuyler F. Heim * Mile 4.5

The Commodore Schuyler F. Heim Bridge was built by United Concrete Pipe Corporation for the U.S. Navy. The impressive structure, one of the largest vertical lift bridges in the country, was designed by State of California bridge engineers.

On January 10, 1948, upwards of one thousand people showed up for the bridge dedication and transfer of proprietorship from the U.S.Navy to the Port of Long Beach. In 1964 the state upgraded the Terminal Island Freeway (State Route 103) and took over the bridge as part of the California highway system.

Built at a cost of $5,331,500, the vertical lift bridge replaced the temporary swing bridge built by the Navy a few years earlier. It is named for Commodore Schuyler F. Heim, the first commanding officer of the Terminal Island Naval Base.

This aerial view of the Badger Avenue and Schuyler Heim bridges was taken in 1953. The view is looking north across Cerritos Channel toward the Ford plant and oil fields. Photo: CALTRANS.

The 750–ton, 240–foot lift span is raised by two 100 h.p. electric motors mounted atop each of the 272–foot lift towers, the nation's tallest in 1948. Vertical clearance in the full open position is 167 feet. Even with nearly 50 feet of clearance when closed, this bridge opens for water traffic upwards of 8,500 times a year; the busiest drawbridge in the state.

The close proximity of the two bridges has caused numerous accidents over the years; some serious. According to one Coast Guard official, a lack of communication between the operators of the two spans is a big part of the problem. Another is the age of the structures and their physical condition.

During a short period in 1978 there were three occasions when one or the other of the bridges failed to open when requested. In one, the failure contributed to the collision between a pleasure boat and a tug with tow. The collision resulted in the sinking of the pleasure craft.

In 1988 the aging Heim bridge received a $2,000,000 refurbishing. During the five months the span was closed, steel grating replaced the pitted and patched roadway and the entire electrical control system was updated. The operator's control room was doubled in size and equipped with computerized controls that allow the leaf to be raised and lowered with the push of a single button. Damage caused by the 1987 earthquake was also repaired.

One of the many signaling devices provided for safety on the bridge is an automatic alarm system operated by a sensor to warn drivers if their vehicle exceeds the 15–foot height limit. When the sensor detects a high load, a siren sounds, a sign warning of the high load is illuminated, and warning blinkers are changed from yellow to red. Unfortunately, some drivers ignore all the signals. Flatbed trucks, stacked high with scrap autos heading for the steel scrap yards on the island often strike the overhead beams, scattering fenders, hoods and other auto parts onto the roadway.

In the 1950s, during the height of the oil subsidence problem, the lift towers sank slightly causing them to tilt 1-1/2 inches to the east. Until corrected, there was fear the lift span might bind on the way up or down. An article relating to the problem appeared in a local paper under the heading, "The Leaning Towers of Long Beach."

This bridge is tended 24 hours daily.

NEWPORT, ANAHEIM AND ALAMITOS BAYS HUNTINGTON HARBOR, SANTA ANA, AND SAN GABRIEL RIVERS

Beginning around the turn of the century, several bridges were built across the arms of several coastal waterways. The structures were built by Orange and Los Angeles Counties, the State of California and the Pacific Electric Railway Company. Most of the state bridges were built along the Pacific Coast Highway, beginning about 1910.

While none of the structures were moveable bridges, nearly all of the earliest ones had *removable spans*. How many of those spans were ever removed to allow the passage of water craft is not known, but it was probably not many.

One such opening did occur in June of 1924, however. The span of a state owned bridge across part of Alamitos Bay at Coast Boulevard, was removed to allow a large dredge to pass through. It most likely opened again to let the dredge back out, but that event is not chronicled.

The action of the tide was used to lift, and later replace the 168–ton span. The feat was accomplished by placing a barge under the span at low water, then letting the incoming tide raise the span free of its pier. Once the dredge had passed through, the span was floated back into position and lowered with the ebb tide. With careful planning traffic was tied up for only five hours.

On March 10, 1933, the disastrous Long Beach earthquake damaged several bridges from Long Beach to Newport Beach. One of the most seriously damaged structures was the state removable span bridge across Anaheim Bay on the Pacific Coast Highway. The roadway gave way at the expansion joints at each end and somehow managed to leave the bridge nine inches shorter than before the quake.

GOLETA SLOUGH

Goleta

Until 1862, a natural harbor existed at a site that extended north of the area encompassed by Hollister Avenue, beyond Stork Road and east of the Santa Barbara Airport. The harbor sheltered large sailing vessels that came to take on cargoes or to make repairs. Some shipbuilding also was carried on in the harbor.

In 1862, however, a great storm caused the harbor to fill with silt. Ships were no longer able to navigate Goleta Slough to what was left of the harbor.

Lumber schooners called at Goleta Landing in the late 1800s and on into the early part of this century. Standing offshore, they sent lumber to the landing and picked up farm products and asphalt brought down the slough on boats and barges.

On December 5, 1929, the Santa Barbara County Board of Forestry completed a bascule bridge across the slough, west of its junction with Atascadero Creek. The structure had 21 wooden trestle spans and one 19–foot bascule span. There is no record of how often the bridge opened for water traffic, but according to local historian and author, Justin Rohge, oil company service boats were common on the slough into the 1930s. He remembers seeing the bridge lift to let the boats through.

On February 2, 1942, a Japanese submarine surfaced off the Santa Barbara coast and fired approximately 25 shots at an oil refinery located not far from the drawbridge. A small amount of damage was done to one building, but that was all.

The bridge site later became part of a U. S. Marine Corps base and the bridge was replaced with a fixed structure.

END

BIBLIOGRAPHY AND SOURCES

This listing applies to all Chapters of this book.

California Drawbridge Regulations 1967, 1984, 1986, 1994; *California Highways* 1917-1923; *California Highways and Public Works* 1924-1960; *Going places* 1960-1994; *Construction News* 1936-1978; *Overland Monthly* 1907-1958; *Pacific Constructor* 1932-1937; *Western Engineering News & American Railway Journal* 1907-1914; *U. S. Army Corps of Engineers Annual Reports from* 1875 *on; U. S. Army Corps of Engineers Port Series; U. S. Coast and Geodetic Charts;* U. S. Coast Guard *List of Bridges over the Navigable Waters of the United States* 1925,1927, 1935, 1941, 1948; *United States Coast Pilot, Pacific Coast,* 1903-1968.

Bancroft Library; California State Archives; California State Library; California State Department of Transportation (CALTRANS); National Archives-Pacific Sierra Region; Southern Pacific Railroad.

The following is in addition to the above and applies to the individual chapters as listed.

Chapter One

Alta California 8-30-1849, 7-1-1951; *The Bay of San Francisco,* by Lewis Publishing Company; *Beginning of San Francisco,* by Zoeth Skinner Aldridge; *Bulletin* 12-2-1855; *Crofutt's New Overland Tourist and Pacific Coast Guide for* 1880; *Historic Evaluation of the Third Street Bridge,* by the City and County of San Francisco Public Works Department 1985; *Sacramento Transcript* 11-5-1850, 8-5-1859; *The San Francisco Bay Area ,* by Mel Scott; *San Francisco Municipal records (annual reports)* 1879 *through* 1917; *San Francisco Public Works Department Annual Reports; Vanished Waters,* by Nancy Olmsted 1986.

California Historical Society; San Francisco Public Library, Special Collections; U. S. Congressional Record.

Chapter Two

Daily Encinal 12-22-02; *History of Alameda County,* by F. Clinton Merritt, volumes I and II, 1928; *History of Contra Costa County California,* by W. A. Slocum & Company 1882; Undated clippings from the *Oakland Tribune, Daily Encinal* and *San Francisco Chronicle*; *Ye Olden Oakland Days* 7-14-20, 10-22-22.

Alameda Historical Society; Alameda Library; Alameda County Public Works Department; Contra Costa County Library; Oakland Library; Oakland Public Works Department; San Francisco Public Library, Special Collections.

Chapter Three

Alta California 8-30-1849; *Argus* 10-7-69; *A Centennial,* by, Bruce A. MacGregor and Richard Truesdale 1982; *The Centennial History of Newark,* by Bruce A. MacGregor 1976; *Drawbridge, California: A Hand-me-Down History*, by O. L. "Monty" Dewey; Dumbarton Bridge Company brochure, 1927; *History of San Jose and Surroundings,* by Fredrick Hall 1871; *Motorland* September 1927; *Newark Argus* 9-30-69, 10-7-69, 10-22-71, 9-20-84, 8-30-85; *San Mateo Times* 2-5-79, 11-29-91, 12-6-91; *South Pacific Coast,* by Bruce A. MacGregor and Richard Truesdale; *Western Railroader* June 1968.

California State Toll Bridge Authority; Port of Redwood City; Redwood City Library; San Francisco Bay and Wildlife Society; San Francisco Public Library, Special Collections; San Mateo Library; Union Pacific Railroad.

Chapter Four

Chronological History of Marin County, Volume II, by C. O. Meret; *Glimpses of Belvedere and Tiburon,* by Landmarks Society 1964; *History of Contra Costa County* 1917; *History of Corte Madera,* by Christine Cadero; *History of Napa*, by Tillie Kanaga 1901; *History of Napa County,* by W. F. Wallace 1901; *History of Sonoma County California,* by Ernest L. Finley 1926; *Inventory of the County Archives of California, Napa County* 1941; *Larkspur Past and Present* The Larkspur Heritage Committee 1991; *Mare Island Grapevine* 8-3-79, 4-2-77; *Marin County*, by Alley Bowen S.F. 1880; *Motorland* July 1966, October 1931, September 1970; *Moveable Bridges*, by Hovey; *Napa County Record* 4-23-69; *Napa Register* 5-16-78, 5-17-79; *The Northwestern Pacific Railroad,* by Fred A. Stindt 1964; *Petaluma Argus* 10-16-06; *Pictorial History of Belvedere* Landmarks Society; *San Francisco Examiner* 3-3-29; *Sonoma County,* by Alley Bowen S. F. 1881; *Sunset* November 1902; *Vallejo Times Herald* 7-27-59, 7-30-59, 4-2-64, 12-17-67, 3-28-80, 3-29-80, 5-1-80; *VIP* 3-27-80.

Petaluma Museum Association; Petaluma Public Works Department; Dudley Key, Port Townsend, WA.; Marin County Public Works Department.

Chapter Five

Colusa County, by Rogers; *Corning Observer* 6-23-49; *Grizzly Bear* 5-19-24; *History of Butte County,* by George C. Mansfield 1918; *History of Butte County,* by Joseph F. McGie; *History of Colusa and Glenn Counties*, by Charles McCormish 1918; *History of Sacramento County,* by G. Walter Reed 1923; *My Playhouse was a Concord Coach,* by Mae Helene Boggs 1854; *Red Bluff Daily News* April 1956; *Sacramento Transcript* 8-5-1850, 11-5-1850 and other undated issues; *Steamboats in the Valley,* by Fred S. Cook 1971.

Community Memorial Museum of Sutter County; Sacramento County Department of Public Works; Colusa County Public Works Department.

Chapter Six

History of Sacramento County, by G. Walter Reed 1923; *History of Solano and Napa Counties California,* by Hunt, Marguerite and Harry Lawrence Gunn; *Sacramento Transcript* 8–5–1850, 11–5–1850, various undated issues; *Steamboats in the Valley,* by Fred S. Cook 1971.

Rio Vista Museum Association; Richard Brown, Bakersfield; Sacramento Department of Public Works.

Chapter Seven

Annals of Stanislaus, by Jack Brotherton; *Bancroft's Guide to Travel by Rail-Stage and Steam Navigation in the Pacific States* 1871; *History of Fresno 1919; History of Merced County; History of Stanislaus County* 1880; *History of Stanislaus County,* by George H. Tinkham 1921; *Merced County Sun* 6–16–11, 6–23–11; *Merced Express* 2–6–1886; *Minutes of Merced County Supervisor's Meetings* 1880–1940; *Stockton Journal* 1852, 1861, 1862.

Port of Stockton; San Joaquin County Department of Public Works; San Joaquin County Library; Santa Fe Railroad; Solano County Department of Public Works; Stockton Department of Public Works; U.S. Navy; Western Pacific Railroad.

Chapter Eight

Antioch Library; Contra Costa County Library; San Joaquin County Library; Hartmann & Setness, Stockton; San Joaquin County Public Works Department; Santa Fe Railroad; Reclamation District 2027; Reclamation District 2030.

Chapter Nine

Eureka library; The Northwestern Pacific Railroad Historical Society; Humboldt County Historical Society; Humboldt County Public Works Department; Hart W. Corbett.

Chapter Ten

Cultural Resources Assessment Henry Ford (Badger Avenue) Bridge, by Portia Lee, Ph.D. July 13, 1993; *Los Angeles Board of Harbor Commissioners Annual Reports 1913-1947; Los Angeles Times* 11–22–08, 10–29–78; *The News Pilot* 6 –13–30, 9–20–89; *Southern California's First Railroad,* by John W. Robinson.

Goleta Valley Historical Society; Newport Beach Public Library; Port of Long Beach; Port of Los Angeles; Raymond F. Berbower, Assistant Chief Harbor Engineer (retired), Port of Long Beach; San Pedro Bay Historical Society.

INDEX

Alameda 37, 41,44,104
Alamitos Bay 169
Alfred Burel Co. 124
Alvarado 45
Alviso v,45,105
American Bridge Co. 134,147-148,163,166
American Toll Bridge Co. 140-142
Anaheim Bay 169
Andrus Island 112
Antioch 47,103,146
Arcata 154,156-158
Army Point 56
Bacon Island iv, 119,144-147
Bahia 49
"Banana Boats" 9
Barrett & Hilp Const. Co. 15
Bay Area Rapid Transit (BART) 52,54
Bay Farm Island 22,32-33,
Belvedere 73
Belvedere Cove 73-74
Ben C. Gerwick Construction Co. 54,66
Benicia 47,52,61,65,85
Benicia-Martinez Ferry 142
Benson's Ferry 106
Bidwell, General John 83
Bidwell Landing 83
Bishop Cut 152
Black Point 64-65
Borden Highway 135,144,146,148
Brannan Island 101
Bridges (description) iii-iv,vi
Bridges (names/locations)
"A" St., Redwood City 35
Alhambra Creek (S.P.) 49
Alice St., Oakland 24,
Antioch 140-142
"B" St., Redwood City 35
Bacon Island 144-146,150
Badger Avenue, Long Beach 165-168
Bay Farm Island 32-33
Bay Farm Island (bike) 34
Bel Marin Keys 70
Benson's Ferry 106-107
Benson's Ferry (Western Pacific) 107
Bishop Cut 152-153
Black Point 64,65
Brandt's Ferry 132
Brannan St., San Francisco 8
Brazos (S.P.) 53-55,60
Butte City 84-85
Cerritos Channel (U.P.) 165
Cerritos Channel (U.S.Navy temp.) 165
Chipps Island (Sac.Northern) 49
Colusa 85-86
Commodore Schuyler Hiem 163,165-168
Connection Slough 150
Cordelia Slough (S.P.) 49
Corte Madera Creek 74
Corte Madera Creek (N.W.P.) 76
Coyote Creek 43
Coyote Hills Slough 45
Crows Landing 125-126
"D" St., Petaluma 68
Daggett Road 135
Davidson, Jack 141
Dickenson Ferry 123
Dumbarton (S.P.) 37-38
Dumbarton (hwy.) 38-40
Durham Ferry 128-129
East Bradford, Redwood City 36
Edison St., Stockton 138
Elk River Slough 159
Eureka Slough (No. Coast R.R.) 55-156
Eureka Slough (U.S.101) 156-157
Eureka Slough (Pacific. Lumber) 157
Eureka Slough (N.M.Derby) 157
Firebaugh 122,133
Fourth St., San Francisco) 10-12,16
Francis "Lefty O'Doul (San Francisco) 15
Freeport 97-98,105
Fremont Slough, Stockton 136
Freshwater Slough (P.H.Christianson) 158
Freshwater Slough (M.Brazil) 158
Freshwater Slough (private) 158-159
Fruitvale Ave., Oakland 26-28,32
Gallinas Creek (S.F.& N.P.) 78
Galt New Hope 106
Garwood Ferry 133
Gerald Desmond 160,162-163
Goleta Slough 169
Grand Island 116-119
Grandville (N.W.P.) 64
Grant Line Canal 149-150
Grayson 126-127
Greenbrae 75-76
"H" Street, Sacramento 91
Hamilton City (Gianelli) 82-83
Hamilton City (No. Elect.) 83-84
Harrison St., Oakland 24-25,31
Haystack Landing (N.W.P.) 67
Hell Gate 74
Henry Ford (Badger Ave.) 166
High St., Oakland 27,30-32
Hill's Ferry 121,124,
Honker Cut 153
"I" Street, Sacramento 91,94,105
Islais Creek, San Francisco 17-19
Isleton 99,105
Jacobs #1 135,144
Jacobs #2 135
James Irvine 56
Jefferson Street, Sacramento (U.P.) 94
Jiboom St., Sacramento 23,103-104
Knight's Landing (S.P.) 88-89,105
Knight's Landing (Hwy113) 89
Lathrop (Hwy.120) 132
Lathrop (Manthey Rd.) 130
Lathrop (S.P.) 128-129
Lathrop (W.P.) 130
Lawrence St. (S.P.), Napa 52
Levon Hapag Nishkian 16
Lincoln St., Stockton 137-138
Little Potato Slough 115
Long, San Francisco iv,9-10,17
Los Angeles & Salt Lake R.R. (Long Beach) 160,161
"M" Street, Sacramento (Sac. No.) 92-93
Mad River Slough (N.W.P.) 157
Mad River Slough (State Route. 255) 158
Mad River Slough (Humboldt No.) 158
Mad River Slough (Humboldt Co.) 158
Main St., Stockton 137-139
Mandeville Island 150-151
Mare Island 57-60
Martinez (S.P.) 46
Maxwell 52-53
Meridian (No. Elect.) 86-88,105
Middle River (Victoria Isl...) 146-147
Middle River Station (Santa Fe.) 147
Miller Sweeney 28
Miller's Ferry 108
Mokelumne River 110
Navy Rd., Stockton 135-136
New Hope Landing 108
Newark Slough 43
Nicolaus (Feather River) 105
Nine Mile 127
Novato Creek (S.F.&N.P.) 71,77-78
Novato Creek (Ignacio) 71
Novato Creek (S.V.) 71
Old River (Victoria Isl..) 148
Orwood (Santa Fe.) 147-148
Pacheco Slough (S.P.) 48
Pacheco Slough (Hwy.) 48
Paintersville 98,105
Park St., Oakland 24,27,29-30,31
Peter R. Mahoney, San Francisco 16

Potato Slough 115
Richardson Bay (Redwood Bridge) 72
Richardson Bay (N.P.C.) 73
Rio Vista 100-103,105
Ryer Island (Miner Slough) 117
Sacramento-Yolo 90
San Mateo Hayward 40,42
Santa Fe (Edison St., Stockton 137-138
Santa Fe (Stockton) 134
Santa Rita 123
Seal Slough 42
Sears Point 62
Seventh St., Oakland 21
Seventh St., San Francisco 12
Sixth St., San Francisco 12-13
Snodgrass Slough (S.P.) 113
Sonoma Creek 61
Squaw Hill 82
Strawberry Point (N.P.C.) 73
Sutter Slough 117-118
Tehama (S.P.) 79
Tehama ("wagon") 80-81
Third St., Napa 52
Third St., San Francisco 13-17
Three Mile Slough 94
Tolay Creek (Tubbs Island) 62
Tolay Creek (S.P.T.R.Co.) 62
Tower 93,105
Tracy Rd. 144
Tule Slough 105
Twelfth St., Oakland 20,32
Twin Cities 114
Tyler Island (Eddie's Bridge) 111-112
Tyler Island (S.P.) 112
"U" Street, Sacramento (Memorial) 93
U.S.Navy (pontoon) 161
Vallejo 56-58
Victoria Island 146,148
Walnut Grove 98-99,105
Walnut Grove (Georgiana Sl.) 112
Warm Springs Slough (Mud Slough) 43
Washington St., Petaluma 69
Washington St., Stockton 137-139
Webster St., Oakland 21-25 27,32,104
West Basin (S.P.) 163-164
Whitmore's Ferry 127
Williams 144
Wingo, Napa Slough 62
Wingo, Sonoma Creek (N.W.P.) 62
Wingo, Steamboat Slough (S .V.) 62
Woodson 81-82
Zuckerman Bros. 151-152
Brodrick 92
Brown, Richard E. 101
Burns Cutoff 135
Butte City 84-85
Butte County 82
Byersville 126
California Bridge Co. 31
California Dept. of Fish & Game 109
California Toll Bridge Authority vi
CALTRANS 79-80,103,146
Cargill (Leslie Salt Co.) 45,56
Carquinez Strait 46,140,142
Cerritos Channel 165-169
Channel Street 8-19
Chester 123
Chico Landing 82-83
China Basin 13
Chipps Island 142
Chowchilla 123
Clark and Kerney Co. 149
Clinton v
Colusa 82, 85-86,88
Colusa County Museum 88
Concord 48
Contra Costa County 20,25,46-48,140,148
Corinthian Inlet 73
Corning 81
Corte Madera 74,76,129
Courtland 98,116-117
Coyote Hills Alameda Co. Flood Control Channel 45
Creeks
Alhambra 48-49
Atascadero 170
Clear 79
Corte Madera 74-75
Coyote 43-45
DeVrees 17
Gallinas 78
Islais 17-19
Mission 8-19
Novato 69-71
Pacheco 47
Petaluma 66
Redwood 35
San Antonio v,20,27
San Rafael 77
Sonoma 60-61
Tolay 60,62-63
Walnut 47
Crenshaw, Hiram 82
Crows Landing 125-126,129
Darby Langdon & Co. 25
De Silva Island 72
Delta Farms 150-151
Derby, N.M. 157
Drawbridge 45
Dumbarton Bridge Co. 38
Dumbarton Landing 37
Dumbarton Point 38
Duncanson Harelson Co. 23,33
Dutton's Landing 53
Egg Boats 66,68
Eight Mile Rd. 152-153
Eldorado St., Stockton 136
Empire City 121,127
Empire Island 153
Erickson, Phillips & Weinberg 52,149
Eureka 154-159
Fabian & Bell Canal 149
Fairhaven 154
Firebaugh, Andrew 122
Fish Camp 144
Ford Avenue 165-166
Ford Motor Co. 165-166
Franklin Road 106
Freeport 97,105
French Camp 132,143
Fresno 121,134
Galerie de Blanche 11
Galt-New Hope 106
Gilman, T.C. 20
Glenn County 82,85
Golden Gate Bridge Dist. 76
Golden Gate Ferries Inc. 60,140
Goleta 169-170
Grand Island 116,145
Grandville 64
Grant Line Canal 149
Graves, Roy 11
Grayson 126-127,133
Green's Landing 113
Greenbrae Junction 75
Guy F. Atkinson Co. 161,163
Hamilton Air Force Base 70
Hamilton City 82,84
Harbor Tug and Barge Co. 29,54,179
Haystack Landing 67
Herndon 121
Hill's Ferry v,121,124-125,129,133
Hollister Ave., Goleta 169
Hood 117
Howard Road 143-144
Humboldt Bay iv 154,156
Hunters Point 9
Ignacio 70
Imola Ave., Napa 52
Isleton 92,99,103,110,112,114,116
Kaiser Steel Co. 52
Kashow, Isreal 73-74
Kentfield 74
Kentucky St., San Francisco 17-18
Key, Dudley 76
King Island 152-153
Knights Landing 88
Laguna Dolores 8
Lake McLeod 136
Lake Merritt 20

Larkspur 73
Llewellyn Iron Works 160
Long Beach 160-161,163,166-167,169
Long Beach Naval Station & Shipyard 167
Lord and Bishop Const. Co. 109
Los Angeles 164,169
Los Banos 122-123
Lower Jones Tract 119,144
Manzanita 72
Mare Island 51,56-57,59-60
Mare Island Strait 52
Marin County 64,72,74,129
Martinez 46,48,65,142,146
Marysville 105
Marysville-Shasta Stage 84-85
Maywood Colony 81
McDonald Island 151
Menlo Park 41
Merced County 123-124
Meridian v, 86-88,105
Merritt Island 117
Miller & Lux 122
Miller's Ferry 108
Mission Bay 8-9
Mission Creek Conservatory 12
Modesto 126-127
Moore Drydock Co. 97
Mossdale Wye 129-132
Most Beautiful Moveable Bridge Award 93,102,112-113
Murietta, Joaquin 124
Murphy Pacific Co. 42
Napa County 49,52,64
Napa (City) 49-51,60-61,65
New Hope Landing 108
Newark 37
Newman 124
Newport Bay 169
Newport Beach 169
Nicolaus 104-105
Nuts and Bolts case 27-28
Oakland v,41,46,65,104,147
Oakland/Alameda 20-34
Oakland Estuary iv,21-33,104
O'Connell, Kevin 12
Orange County 169
Oxford 105
Pacheco v,47-48
Pacheco Pass 122
Pacific Bridge Co. 126
Pacific Coast Highway 169
Pacific Lumber Co. 157
Paradise Cut 149
Petaluma 51,65-66,68-69,71,85
Petaluma Landing 67
Peterson, Eddie 111
Pine Point 73
Pittsburg 142
Port Costa v,47
Port of Long Beach 163,166-167
Port of Los Angeles 166-167
Port of Redwood City 35
Port of San Francisco 16
Port of Stockton 135-135
Posey Tube 23,25
Princeton 85
Prospect Island 117-118
Railroad St., San Francisco 13,19
Railroads
Alameda & San Joaquin 130
Bayview & Potrero 9
California Northern 61-62
California Pacific 90
Central Pacific 97
Eureka & Klamath 158
Eureka Southern 155-156
Freeport 97
Harbor Belt Line 164,166
Los Angeles & Salt Lake 160-161
North Coast 155
North Pacific Coast 73
Northwestern Pacific 61,64-65.71-72,76,78,155,157
Northern Electric 83-84,86
Oakland Antioch & Eastern 49,142
Pacific Electric 160,164,169
Port of Stockton 136
Sacramento Northern 49,92-93,105,113-114
Sacramento Valley 97
San Francisco Oakland 21
San Francisco & North Pacific 67
San Francisco & San Joaquin 147-148
San Joaquin Valley 129
San Pedro, Los Angeles & Salt Lake 160
Santa Clara Valley 43
Santa Fe 10,13,18,37,134,136,138,147-148
Santa Rosa & Carquinez 53
Sonoma Valley 62,71
South Pacific Coast 21,43-44
Southern Pacific 10,13,16-18,24-25,27,32,39,43-49,52,55,79,82,84, 88-89,91-92,112,128-131,136-137,163-164
State Belt Line 15-16
Tiburon & San Rafael 76
Union Pacific 49,94-95,160,165,167
Vallejo, Benecia & Napa Valley Electric 52
Western Pacific 17,21,37,107,115,130,136,140
Rattlesnake Island 160
Reclamation District 2027 150
Reclamation District 2030 151
Red Bluff 79
Redding 79,35,37
Redwood City 35,37
Richards, Harry 76
Richardson Bay 72
Richmond 65
Rio Vista 95,100-103,117-119,145
Rivers
American 79,103-104
Coyote 43
Feather 104-105
Merced 121,125
Middle iv,118,143-147,149
Mokelumne 106,110,113,115
Napa v, 49,57,59
Old 147-149
Petaluma v, 64-69
Sacramento v,46,66,79-105,110-113,116-117,119
San Gabriel 169
San Joaquin v,46,66,106,110,119,121-142,144,147-149
Santa Ana 169
Tuolumne 120-121,127
Roberts Island 132-133,143-144
Rohge, Justin 170
Roosevelt Base 161
Ross 74
Ross Construction Co. 166
Ross Landing 74
Rough & Ready Island 134-135,144
Russell's Ferry 126
Ryer Island 117-118
Sacramento 61,79,85,89,98,106,112,118,142,144 91,97,100,114,116,140
Sacramento River State Park 86
Sacramento Ship Canal 118
Sakata Bros. (Prospect Farms) 117
Samoa 154
San Anselmo 73
San Francisco 8,19,26,35,41,51,65,66,70,74,77,79, 85,121-122,126,129,154,167
San Francisco Bay iv-v 35,37-38,52,61,72,142
San Francisco Bay Wildlife Service 45
San Joaquin City 128-129
San Joaquin County 116,130,144,146,148-150
San Jose 41,45
San Leandro Bay 20,27,32,34
San Mateo 41-42
San Mateo County 35,42
San Pablo Bay 60,69

San Pedro 164
San Quentin 74,76
San Rafael 71-72,77
San Rafael Ship Canal 77
Santa Barbara 169-170
Santa Cruz 45
Sausalito 65,73
Sears Point 60,71
Sears Point Toll Road 56,60-63
Shasta Dam 89
Shellville 61,63
Sherman Island 103,141
Shilo Rd. (Modesto) 127
Ships, boats and barges
A.C.Freese 126
Arrow 51
Belle 79,90
Centennial 121
Colusa 89
Comanche 79
Crockett 67
Daisy Grey 134
Dana 88
Delta King 79,94
Delta Queen 79,94
S/S Diane Dollar 23
Dolphin 51
Ellen 28
Frisby 51
Frolic 36
Gem 79
S /S General Hubbard 160
Georgia 61
Gold 51
Ida 74
Ilice 102-103
James M.Donahue 76
J.R.McDonald 121,123
Juanita 21
S /S Kaimain 141
La Bonita 32
La Placentia 53
S/S Landcaster 22-23
Margaret 67
Montezuma 67
Napa City 51
Napa Valley 51
Ohio 98
M/V Pasadena 141
Perseverance 35
Petaluma Queen 68
Placer 89
Ramon 50
Saint Helena 51
San Joaquin II 98
Sea Lion 23
Serrana 98
Solano 47
Sutter 51
Tamalpias 25,74
Thoroughfare 38
Tuolumne City 127
Twilight 77
Tyee 120
Walker 77
S/S Washington Bear 141
Zinfandel 51
Sloughs
Asylum 60
Cache 95,118
Connection 150
Cordelia 49
Dutchman 56
Elk River 159
Eureka 154-157
Fremont 136
Freshwater 158
Fresno 121
Gallinas 77-78
Georgiana 110,112
Goleta 169
Laird 126
Little potato 115-116
Mad River 157
Midshipman 62
Miner 117-118
Montezuma 49
Mormon 134-137,139
Mud 43-44
Napa 61-62
Newark 43
Norfolk 61
Pacheco 47-48
Potato 115
Seal 42
Snodgrass 113-114
South 56
Steamboat 62,98,116-119,145
Stockton 136
Suisun 49
Sutter 117
Three-Mile 94,110,117,119-120,145
Tule 105
Warm Springs 43-44
Smith Canal 13,140
Solano County 46-47,52,60-61,64
Sonoma County 61,64,100
Sonoma Embarcadero 61
Squaw Hill 81
Stanislaus County 124-126
State Harbor Commission 16,18
Staten Island 108
Station Island 43-45
Stevenson. J.J. 125
Stinson Beach 72
Stockton 47,85,106,121,127,130,132-134,137-139,144,146-149,152
Stockton Ship Canal 133-136,140
Strauss Bascule Bridge Company 98,163,166
Suisun Bay 142
Suisun City 49
Suscol Landing 49
Sutter County Museum 88
Sutter Island 117
Sweetzer Landing 71
Tamalpias 72
Tehama 79,81,88
Terminal Island 160,165-167
Terminal Island Federal Prison 167
Terminal Island Naval Base 167
Terminous 115
Third St., San Francisco 13-19
Thomas Rigging Co. 29
Thompson Bridge Co. 10
Thornton 106
Tiburon 73
Tracy 129-130,144,146,149-150
Tule Jake's Road 95
Tuolumne City 127
Turner Cut 151
Tyler Island 108,110-112
Twitchell Island Ferry 111
United Concrete Pipe Corp. 167
United Fruit Co. 15
Union Island 143-144,149
U.S. Army Corps of Engineers iv,20,22,26,48-50,52,56,65-66,72,93,95,140,157,159,163
U.S. Bureau of Reclamation 96
U.S.Coast Guard iv,vi,86,95,156,164,168
U.S.Marine Corps 170
U.S.Navy 57-59,62,135,161-162,165,167
Vallejo 51,64,65
Victoria Island 146-148
Walnut Grove 98,110,112-114,117
Washington Lake 95
Washington St., Petaluma 67
West Basin 163-164
Wilmington 164-166
Winchester Lake 105
Wood Island 100
Woodbridge 106
Woodward Island 147
Yolo Barge Canal 94-95
Yolo Causeway 95-97
Yolo County 91,97
Yosemite Lake 140
Yuba City 105
Zuckerman, Maurice & Roscoe 151

AUTHOR PROFILE

Bernard Winn grew up with a love for all things maritime. Born at the southern edge of San Francisco, it was a short walk from his father's factory near Sixth and Brannan Streets to Mission Creek, and the busy piers along the waterfront. There he fished and watched the unloading of foreign ships and dreamed of the day when he would sail to exotic places. While still in high school, his interest in the sea took him to evening classes in celestial navigation at the Ferry Building.

To satisfy his natural wanderlust, he spent the summers of 1937-1939 as a hobo, riding freight trains across America.

In 1940, at the age of 20, Bernard's dream of going to sea became a reality. Shipping as officer's messboy on the Swedish tramp steamer Herma Gorthon, he embarked on a long voyage to several ports in the Orient, Dutch East Indies and Mexico.

Following the call of the sea, he joined, in turn, the Scandinavian Seaman's Club, Sailors' Union of the Pacific, Seafarers International Union, National Organization Masters, Mates & Pilots of America and later, the Inlandboatman's Union of the Pacific.

During the war he rose to the position of Second Mate and held the rank of Lieutenant (jg) in the U.S. Maritime Service and Ensign in the U.S. Naval Reserves.

After the war, he earned a BA degree from the California State University at Hayward.

Writing credits include four books on European travel, primarily via unusual railways, and his most recent book, published in 1993, titled *Arch Rivals: 90 Years of Welcome Arches in Small-Town America.*

Now 75 and retired, Bernard and his wife, Audrey, live on the Embarcadero in San Francisco, a stone's throw from the Third Street Bridge where his love for the sea was nurtured.

His early fascination with drawbridges, which fostered this book, was rekindled in the 1960s, while working on tugboats and ferries for Crowley Tug and Barge Company.